Leadership and Sustainability in the Built Environment

Leadership and sustainability have separately been the subject of numerous studies in a built environment context over the years, but they have yet to be addressed together. The real impact of legislation and guidelines designed to promote sustainability within the construction industry is closely linked to the leadership behind it, as this book explores in a variety of ways.

Featuring research from five different continents, the international scope of this book allows a comparison of experiences in different types of economies and cultures. The interdisciplinarity of this subject is also reflected in the backgrounds of the contributors, with a significant number of perspectives derived from business and management research.

The issues examined in this book are essential reading for all researchers, decision-makers and graduate students in the built environment.

Alex Opoku recently took up a position as a Senior Lecturer in Quantity Surveying at London South Bank University, having previously conducted research in the Department of Engineering at Cambridge University.

Vian Ahmed is a Professor and Director of Postgraduate Research Studies at the Built Environment department of the University of Salford.

Spon Research

Publishes a stream of advanced books for built environment researchers and professionals from one of the world's leading publishers. The ISSN for the Spon Research programme is ISSN 1940-7653 and the ISSN for the Spon Research E-book programme is ISSN 1940-8005

Published:

Free-Standing Tension Structures: From tensegrity systems to cable-strut systems
978-0-415-33595-9
W. B. Bing

Performance-Based Optimization of Structures: Theory and applications
978-0-415-33594-2
Q.Q. Liang

Microstructure of Smectite Clays & Engineering Performance
978-0-415-36863-6
R. Pusch & R. Yong

Procurement in the Construction Industry: The impact and cost of alternative market and supply processes
978-0-415-39560-1
W. Hughes et al.

Communication in Construction Teams
978-0-415-36619-9
S. Emmitt & C. Gorse

Concurrent Engineering in Construction Projects
978-0-415-39488-8
C. Anumba, J. Kamara & A.-F. Cutting-Decelle

People and Culture in Construction
978-0-415-34870-6
A. Dainty, S. Green & B. Bagilhole

Very Large Floating Structures
978-0-415-41953-6
C.M. Wang, E. Watanabe & T. Utsunomiya

Tropical Urban Heat Islands: Climate, Buildings and Greenery
978-0-415-41104-2
N.H. Wong & C. Yu

Innovation in Small Construction Firms
978-0-415-39390-4
P. Barrett, M. Sexton & A. Lee

Construction Supply Chain Economics
978-0-415-40971-1
K. London

Employee Resourcing in the Construction Industry
978-0-415-37163-6
A. Raiden, A. Dainty & R. Neale

Managing Knowledge in the Construction Industry
978-0-415-46344-7
A. Styhre

Collaborative Information Management in Construction
978-0-415-48422-0
G. Shen, A. Baldwin & P. Brandon

Containment of High Level Radioactive and Hazardous Solid Wastes with Clay Barriers
978-0-415-45820-7
R.N. Yong, R. Pusch & M. Nakano

Performance Improvement in Construction Management
978-0-415-54598-3
B. Atkin & J. Borgbrant

Organisational Culture in the Construction Industry
978-0-415-42594-0
V. Coffey

Relational Contracting for Construction Excellence: Principles, Practices and Partnering
978-0-415-46669-1
A. Chan, D. Chan & J. Yeung

Soil Consolidation Analysis
978-0-415-67502-4
J.H. Yin & G. Zhu

OHS Electronic Management Systems for Construction
978-0-415-55371-1
I. Kamardeen

FRP-Strengthened Metallic Structures
978-0-415-46821-3
X. L. Zhao

Leadership and Sustainability in the Built Environment
978-1-138-77842-9
A. Opoku & V. Ahmed

The Soft Power of Construction Contracting Organisations
978-1-138-80528-6
S.O. Cheung, P.S.P. Wong & T.W. Yiu

Fall Prevention Through Design in Construction: The Benefits of Mobile Construction
978-1-138-77915-0
I. Kamardeen

Leadership and Sustainability in the Built Environment

Edited by Alex Opoku
and Vian Ahmed

Routledge
Taylor & Francis Group
LONDON AND NEW YORK

First published 2015
by Routledge
2 Park Square, Milton Park, Abingdon, Oxon OX14 4RN

and by Routledge
711 Third Avenue, New York, NY 10017

Routledge is an imprint of the Taylor & Francis Group, an informa business

British Library Cataloguing-in-Publication Data
A catalogue record for this book is available from the British Library

Library of Congress Cataloging in Publication Data
Leadership and sustainability in the built environment/edited by Alex Opoku and Vian Ahmed.
pages cm
1. Sustainable construction. 2. Sustainable architecture.
3. Leadership. I. Opoku, Alex, editor. II. Ahmed, Vian, editor.
TH880.L43 2014
690.068′4—dc23
2014027244

ISBN: 978-0-415-62242-4 (hbk)
ISBN: 978-1-315-84953-9 (ebk)

Typeset in Sabon LT Std
by Swales & Willis Ltd, Exeter, Devon, UK

MIX
Paper from responsible sources
FSC www.fsc.org FSC® C013604

Printed and bound by CPI Group (UK) Ltd, Croydon, CR0 4YY

Contents

Figures

Tables

Contributors

Vian Ahmed is a Professor in the Built Environment. Her teaching expertise is in construction management and IT at the undergraduate and postgraduate levels. She has broad research interests covering skills, culture and people in construction, including e-learning in construction, construction IT for large organisations and SMEs, ontology development for construction education, construction management and organisation readiness. She is a senior fellow of the Higher Education Academy, with well over 100 journal and conference publications, 18 graduated doctoral students and a similar number of continuing students. She holds a BEng (Civil), MSc (Construction) and PhD in Computer Aided Learning in Construction.

Leighton A. Ellis is an Assistant Lecturer in the Department of Civil and Environmental Engineering at the University of the West Indies, St. Augustine, Trinidad and Tobago. He is a member of the Association of Professional Engineers of Trinidad and Tobago (APETT). He holds both a BSc and MSc (Eng) in Construction Engineering & Management and is presently pursuing a PhD in Construction Management. Ellis' research is in the area of leadership assessment of graduate students using innovative techniques for student assessment.

Fidelis A. Emuze is presently a Senior Lecturer and Department Head, Department of Built Environment at the Central University of Technology, Free State, South Africa. He holds a National Diploma and Higher National Diploma in Civil Engineering, an MSc in the Built Environment, and a PhD in Construction Management. Dr. Emuze has taught and assessed undergraduate and postgraduate courses in Construction Management, Research Methodology and other topics. His completed and on-going research pertains to lean construction and sustainable construction. Dr. Emuze is on the editorial review board of international journals and scientific committees of conferences in the construction research domain.

Chris Fortune has recently joined the Business and Management School at Glyndŵr University, Wrexham, Wales as a Professor of Project Management, having previously held professorial posts in construction project management at Heriot Watt University, Edinburgh, Scotland and at the University of Salford. He is a former Associate Head of School of the Built Environment (Teaching and Learning) and an Associate Dean (Academic) for a Faculty of Business, Law and Built Environment. Professor Fortune entered academe after working as a chartered surveyor and chartered builder in the construction industry for a number of years and possesses an MBA from the Institute of Education, University of London and a PhD from Heriot Watt University. He has a well-established academic track record in teaching and researching issues in procurement and supply chain management, with over 120 publications in conference proceedings and journal papers to his name. Professor Fortune has over 20 years' experience of teaching modules related to procurement practice and theory and supervising dissertations related to sustainable practices, procurement and supply chain management at the undergraduate and post graduate levels.

Yamuna Kaluarachchi is a Reader and Deputy Director of the Research Centre, School of Surveying & Planning, Kingston University. Kaluarachchi is a chartered architect by profession. Her expertise and research interests include sustainable social housing refurbishment, stakeholder engagement, changing behaviour to enhance low-carbon living and disaster resilience of communities and their built assets.

Timothy Michael Lewis has recently retired as Professor of Construction Engineering & Management at the University of the West Indies, Trinidad and Tobago. He holds a MEng in Civil Engineering from the University of Liverpool and an MSc in Technological Economics from the University of Sterling as well as his PhD in Civil Engineering from the University of the West Indies. He is a Fellow of the Institution of Civil Engineers (FICE) and the American Society of Civil Engineers (FASCE) as well as of the FAPETT. He has published extensively on construction management and economics topics.

Ciaran McAleenan, a chartered civil engineer, has designed and managed major water and road projects as well as developing HS&E and road design standards. An award winning international speaker, author and a prolific researcher, he has presented at conferences in four continents including World and National Congresses and CIB Commission Conferences. A corporate member of ICE, McAleenan sits on their editorial board for the *Management, Procurement and Law* journal and on their Steering Group for the Construction H&S Register. He is a Professional Member of ASSE, Fellow of the HEA and member of the faculty at the Ulster University.

Philip McAleenan is an MSc graduate of Queens University Belfast. He also holds certificates in law and education, is a Fellow of the Institute of Leadership & Management and is a Professional Member of ASSE. Managing partner at Expert Ease International, he co-developed the Organisation Cultural Maturity Index and the multi-award winning Operation Analysis and Control model. He regularly presents to CSSE and CIB conferences and the World Congresses on OSH. He guest lectures at ECU, QUB, and UUJ where he teaches ethics reasoning to construction undergraduates. He has written extensively on safety, contributing to manuals, professional and academic papers and books in the UK and USA.

Alex Opoku is currently a Senior Lecturer in Quantity Surveying at the School of Built Environment and Architecture, London South Bank University. He is also the Course Director for the MSc Quantity Surveying programme at the School of Built Environment and Architecture. Prior to joining the London South Bank University in 2013, Opoku worked as Postdoctoral Research Associate at the University of Cambridge, Centre for Sustainable Development in the Department of Engineering. Opoku holds a PhD in Construction & Project Management from the University of Salford. He graduated from Nottingham Trent University with an honours degree in Building Management in 2005 and received an MSc degree in Quantity Surveying Commercial Management from the Leeds Metropolitan University in 2008. Dr. Opoku also has a well-established academic track record in teaching and researching issues in sustainable built environment, with a good number of publications in conference proceedings and journal papers to his name. He believes in making academic research relevant to practitioners and has been engaging with the construction industry through workshops, seminars and conferences.

Andrew K. Petersen is Professor in the Department of Civil Engineering, FH Mainz, University of Applied Science, Mainz, Germany. He has held positions as Senior Lecturer at the University of the West Indies and Principal Lecturer in the Department of Civil Engineering, University of Portsmouth, England. Professor Petersen has introduced a number of new assessment strategies for assessing student learning in the subject of health and safety risk management.

Begum Sertyesilisik is a Visiting Professor in the School of Built Environment at the Liverpool John Moores University and is working in the Department of Architecture at the Istanbul Technical University. Sertyesilisik has been awarded a PhD in the field of construction contracts at the Middle East Technical University. She obtained an MSc in the field of Project Management, an MBA, and a BSC in Architecture at the Istanbul Technical University. She specialises in the areas of project management, sustainability, contract and dispute management. She is a member of the Chamber of Architects in Istanbul and of CIB.

Egemen Sertyesilisik graduated from Istek Kemal Ataturk High School in 2002. He received a BA from the Political Science and Public Administration Department at the Bilkent University in 2007. He then received a MA from the program in Politics and the Mass Media at the University of Liverpool. In 2010 he obtained an MBA at the Yildiz Technical University. Sertyesilisik started his PhD at Marmara University in the field of Political Economy of the Middle East in 2011.

Acknowledgements

We would like to express our gratitude and appreciation to all those who gave us the possibility to complete this book. A special thanks to all the authors who contributed to this book for their time and co-operation in writing these chapters. We would also like to acknowledge with much appreciation the crucial role of the academics who took their time to review the papers.

At a personal level, the editors would each also like to thank their families for their invaluable support during the editing of the book.

Alex Opoku
Vian Ahmed

Introduction

Organizational leadership is vital in the construction industry and a key success factor in the drive towards sustainable built environment. This book showcases high-quality research that examines the link between leadership and sustainability in the built environment. The book consists of chapters underpinned by original research that seeks to shape the future direction in the field of leadership and sustainable built environment and help build a common understanding of the role and contribution of organizational leadership in achieving a sustainable built environment. It brings together these isolated issues serving as a source of reference material for both formal and informal higher education programmes in the built environment, including continuing professional development (CPD) and training programmes.

The book consists of chapter contributions written by built environment experts across the globe. The book is also useful to built environment practitioners in the United Kingdom and internationally due to the current relevance of sustainability in the construction sector. The book consists of nine chapters split into the following four parts:

- Leadership
- Sustainable built environment
- Leadership for sustainable built environment
- International perspective and case studies

The part one covers leadership theories and principles in the context of the built environment whilst the part two examines sustainable development principles and the built environment. Part three explore the link between organizational leadership in achieving a sustainable built environment. The final part provides some international perspective and case studies on the drive towards a more sustainable built environment.

Part I

Organizational leadership

1 Leadership theory and practice for sustainable built environment

Alex Opoku and Vian Ahmed

1.1 Introduction

Leadership is vital in the construction industry and a key success factor in the drive towards sustainability (Ofori and Toor, 2008). Construction organizations need leadership that provides the collective vision, strategy and direction towards society's common goal of a sustainable future. Organizational leaders should embed sustainability approaches in their organizational activities and make sustainable development part of their overall business strategies. The role of leadership in improving the performance and innovation in the construction industry has been receiving increasing attention in recent times (Bossink, 2007). However, less attention has been given to the capability of organizational leadership in promoting construction organizations towards the delivery of sustainable construction projects. The issue of sustainability is growing ever more important and the construction industry has a greater impact on sustainability than any other industrial sector because the construction industry provides benefits to the society as well as causing negative impacts, making it a key sector in the fight for sustainable development (Sev, 2009).

Leaders have an important role in guiding construction organizations toward sustainable practices and it is believed that such leaders require unique leadership styles. Leadership style is all about how people interact with those they seek to lead (Goetsch and Davis, 2006). However, Toor and Ofori (2008) believe that leadership is also about authenticity and not style. There has not been any evidence to show that one particular leadership style is the best (Vecchio, 2002; Giritli and Oraz, 2004). However, Bossink (2007) argues that strategic, charismatic, instrumental and interactive leadership styles influence an organization's innovativeness towards sustainability. Leadership within construction organizations has a role in helping to promote sustainable construction by providing training and awareness on sustainability, developing strategies and producing guidance notes and policies (Opoku and Fortune, 2011). To achieve a sustainable built environment, it is extremely important that organizations are led by leaders who can see beyond just profit to embed the triple bottom line of social, environmental and economic issues in their day to day decisions.

1.2 Leadership – definition and context

Doh (2002) argues that leadership is an executive position in an organization and that it is a process that has influence on others. However, Munshi *et al.* (2005) argue that leaders are essential at all levels of organization and can emerge at different levels within an organization (Newton, 2009). Leadership can be practiced by any individual at different levels within an organization (Riches, 1997) regardless of the position of hierarchy of that individual in the organization (Bass and Riggio, 2006). Leadership is also said to be concerned with the ability of an individual to influence the behaviour of others in order to deal with the desires of the leader (Fellows, Liu and Fong (2003). Ferdig (2007) describes leaders as those who inspire a shared vision, build consensus, provide direction and foster changes in beliefs and actions among followers needed to achieve the goals of an organization. Northouse (2010:3), however, defines leadership as 'a process whereby an individual influences a group of individuals to achieve a common goal'.

From the definitions of leadership above, anyone in an organization could potentially be a leader at some point in time if he or she is involved in a process of influence that involves encouraging and influencing sustainable practices (Taylor, 2008). Organizational leadership seeks to influence other individuals (subordinates, superiors, peers) within the organizations to achieve specific projects, aims and objectives (Gattiker and Carter, 2010). The role of leadership is formally or informally developed within the group and it is the necessary role of the leader to organize, motivate and assign tasks in the group towards achievement. The definition of leadership is continually shifting and difficult to define because it is very much contextual (Bryman, Stephens and Campo, 1996). There are over 350 definitions of the term 'leadership' that have been developed so as to develop the fundamental theory of leadership (Daft, 2005). The most consistent definition of leadership is that of a process of influence and such influence can come from both internal and external stakeholders of the organization (Yukl, 2006). Leadership is now being considered as a process of influencing organizational direction and vision, occurring through the relationships between leaders and their followers (Taylor, Cocklin, Brown and Wilson-Evered, 2011). It is argued that leadership is primarily about influencing individuals to go beyond their selfish short-term interests, to contribute to the long-term performance of the whole group (Northouse, 2010). However, Tabassi and Bakar (2010) add that leadership is not just a process, but a process that involves influences and occurs within a group context, involving personal discovery and development as well as involvement in goal attainment.

1.3 Leadership theories

All of the theories of leadership are correct in one way or other and such theories deal with a leader's move toward the business environment and the followers' opinions of a leader's performance (Northouse, 2010). Some leadership theories centre on the nature of the leader, their personality and traits, whereas other theories centre on identifying the different roles of leaders in terms of what leaders do rather than their characteristics. Some leadership theories also view leadership as specific to the situation, based on the idea that different situations require different leadership styles (Dearlove and Coomber, 2005). Leadership theories identified by Munshi *et al.* (2005) include traits and styles; contingency; transformational/transactional; distributed and structuralist leadership theories. Different schools of leadership theory have evolved over the past several decades such as Leadership traits and style, Leadership behaviour, Contingency approaches, Leader-member exchange (LMX), Great Man leadership theory, Transformational leadership, Charismatic leadership theory and Shared leadership (Bass and Avolio, 1994; Munshi *et al.*, 2005; DeChurch, Hiller, Murase, Doty and Salas, 2010; Yang, Huang, and Wu 2011). A summary of the leadership theories is presented in Table 1.1. The next section discusses the common leadership styles exhibited by leaders in organizations. There is no consensus on the best leadership theory and each theory exhibits certain characteristics expressed in different leadership styles.

Table 1.1 Summary of leadership theories

Leadership Theory	*Characteristics*	*Source*
The Trait Approach	• Individual leader's characteristics or approaches define the trait/style of leadership • Trait leadership theories believe leaders are born and not made	Yukl, 1998
The Behavioural Theory	• Deals with the styles adopted by the leaders for their particular task. • Detail specific behaviours related with effective leadership. • Believe that leaders are made, not born.	Bryman, 1992
The Contingency Theory (Situational theory)	• It is about the appropriateness of different leadership styles in different leadership situations	Burns, 1978
The Leader-member exchange (LMX)	• Believes that leaders form differentiated patterns of relationships with their subordinates • Deals with the direct relationship between leaders and followers	Yukl, 2006
Great Man Leadership	• Believe that great leaders are born to lead; they are extraordinary and exceptional people	Bass *et al.*, 2003
Transformational Leadership	• Transformational is concerned for relationships while Transactional is about concern for process; all aimed at improving leadership outcomes	Bass and Avolio, 1993

1.4 Leadership styles

There have been several studies on how leaders and their styles of leadership promote change (Bryman, 2004) and it is now believed that individual leadership style is a very important factor in innovation (Dess and Picken, 2000). Toor and Ofori (2006) describe leadership style as a combined outcome of the leader's self-related cognitive information, personality traits, the primary motives and thoughts on operating situational variables. It is important that the overall leadership style adopted suit the organization's beliefs, values and assumptions. There are different types of leadership styles, each proving effective depending on the given circumstances, attitude, beliefs, preferences and values of the people involved. Tabassi and Bakar (2010) add that effective leadership style is critical to all successful projects and organizations. Nicolaou-Smokoviti (2004:410) defines leadership style as 'a stable mode of behaviour that the leader uses in his or her effort to increase his or her influence, which constitutes the essence of leadership'.

Many styles of leadership have been proposed for organizational leaders including transactional, transformational, charismatic, democratic, servant, autocratic, consultative, laissez-faire, joint decision-making, authoritative, participative, tyrant, task-oriented, relationship-oriented, production-oriented, employee-oriented, delegating, authority-compliance, impoverished management, team management etc. (Toor and Ofori, 2006). It is suggested that different leadership styles are appropriate in different circumstances and the style of a leader has a major influence on the performance of an organization.

Transformational leadership motivates subordinates to perform beyond the expected levels of performance and can be identified with the goals and the interest of the organization (Bass and Avolio, 1994; Gardner and Avolio, 1998). Transformational leaders lead by example to influence followers' moral, emotional, affective and cognitive behaviour by showing positive qualities and ethics (Bass and Riggio, 2006; Zhu, 2011). Such leaders therefore make decisions that promote ethical policies, procedures and processes in their organizations (Avolio, 2005; Zhu, Riggio, Avolio and Sosik, 2011). Brown and Treviño (2006) cite the seminal work of Burns (1978), who suggests that transformational leaders motivate their followers to think beyond self-interest and work together for a shared cause because of the leaders' moral qualities. It is argued that such qualities exhibited by transformational leaders support and promote innovations in organizations they lead.

Transactional leadership monitors performance and takes the necessary corrective action. Transactional leaders can inculcate moral standards in an organization through effective ethical structures because they have a positive impact on the followers' moral personality (Zhu *et al.*, 2011). However, transactional leadership does not possess the same level of morality as does the transformational leadership (Bass and Steidlmeier, 1999).

Charismatic leaders are very good at shaping the values of others (Brown and Treviño, 2009). They are regarded as visionary leaders who foster good relationships with their followers to achieve excellent performance of the organization's vision through personal characters and behaviours (Hayibor, Agle, Sears, Sonnenfeld and Ward, 2011). The charismatic leadership style communicates vision, energizes others and accelerates innovation processes such as sustainability.

Ethical leaders possess characteristics such as honesty, caring and principles. Ethical leaders communicate with their followers on ethics, set clear ethical standards, use rewards and punishments and make fair and balanced decisions (Brown and Treviño, 2006). According to Riggio *et al.*, (2010), an ethical leader demonstrates prudence, temperance, fortitude and justice in his or her personal characteristics and actions.

Authentic leadership has recently emerged as another form of leadership which complements the work on ethical and transformational leadership (Avolio and Gardner, 2005; Avolio *et al.*, 2004). Authentic leaders are not necessarily transformational, visionary or charismatic leaders (May, Chan, Hodges and Avolio, 2003); however, they incorporate transformational and ethical leadership qualities (Avolio *et al.*, 2004), demonstrate a higher moral ability and are guided by a set of ideals (Lloyd-Walker and Walker, 2011).

Visionary leaders (transformational, charismatic) create a strategic vision of some organizational future to achieve high levels of cohesion, commitment, trust, motivation and hence performance in the new organizational environments (Zhu *et al.*, 2011). Avery (2004) describes visionary leaders as people

Figure 1.1 Word frequency diagram of leadership styles in UK construction industry

Source: Opoku, 2012

who employ a collaborative style for making decisions, share problems with their followers and seek consensus before the leaders make the final decision.

Strategic leadership theory is thought to be similar to the trait theories; it is, however, different as it focuses on individuals at the top of an organization and their effects on strategic processes and results (DeChurch *et al.*, 2010). Strategic leadership style is believed to be the most appropriate leadership style for organizations implementing corporate social responsibility strategies.

Laissez-faire leadership represents a leadership style in which the leader avoids making decisions and using his or her authority and relinquishes responsibility. A laissez-faire leader chooses to avoid taking action and avoid leading. It is believed to be the most passive and ineffective form of leadership (Antonakis, Avolio and Sivasubramanium, 2003). In a study by Opoku (2012) using a mixed method approach involving 15 interviews and 200 surveys, the results showed that transformational and strategic leadership styles influence an organizational change towards sustainability. Figure 1.1 provides the word frequency diagram of the leadership styles among organizational leadership in UK construction organizations, showing the dominance of the strategic style. This was produced from the interview data using the word frequency query tool in NVivo qualitative data analysis software to identify the most frequently occurring words used by interviewees in describing their style of leadership.

1.5 Leadership and the construction industry

Leadership is believed to be an important factor in achieving business success in any organization. Even though the field of leadership is well-researched, Chan and Cooper's (2007) analysis of leaders in the UK construction industry revealed that the understanding of construction industry organizational leadership is to some extent primitive, compared with the rather mature developments of mainstream leadership theories. Research in leadership is increasingly gaining importance in construction management today (Tabassi and Bakar, 2010). The construction industry in general and the UK construction industry in particular is in an era of a difficult socio-economic, cultural, political and business environment. There is an urgent need to promote a positive culture in the construction industry and this requires leaders with positive values and good levels of moral and ethical behaviour to change the conservative paradigm of management in the industry (Toor and Ofori, 2008). There is a clear need for leadership at all levels in the construction industry; at the organizational level of the industry, there is the need for strategic leadership that can develop and improve the industry (Ofori, 2009).

The success of any business organization depends to a great extent on its leadership. There is the need for leaders who can give their organizations a vision, a purpose for collective good and the confidence to innovate (Ofori, 2009). Leadership plays a key role in ensuring success in almost any initiative within an organization. Leadership is all about effective influence on individuals or groups to accomplish an organizational goal or mission (Oke and Gbadura, 2010). Leadership is an important factor in achieving business success (Giritli and Oraz, 2004) and it is critical to the undertaking of any construction project, the management of the enterprise and the development of the industry as a whole. Due to the nature of the construction industry, there is a greater need for effective leadership in this industry than in any other industry. Leadership is critical to the activities of any construction projects as well as the whole industry (Ofori, 2009). There is the need for new leadership paradigms and forms of leadership development in organizations that champion sustainable practices (Taylor, 2008). Toor and Ofori (2008) believe that leadership which lacks ethical behaviour can be treacherous, destructive and dangerous and transformational leadership has the potential to be the most useful to the construction industry.

1.6 Sustainability leadership

Sustainability leaders are those who can respond urgently and effectively to the social, environmental and economic challenges that organizations face today. However, McCann and Holt (2010:209) define sustainability leadership as involving 'a leader concerned with creating current and future profits for an organization while improving the lives of all concerned'.

Organizations therefore need leaders who can define and set visions to achieve success within economic, environmental and social spheres. Lehmann (2008) asserts that there is the need for responsible leadership for business in the twenty-first century. Sustainability leadership will become increasingly important and therefore the next generation of leaders must learn how to integrate their conceptualization of sustainability with the ability to facilitate others in pursuit of that vision. According to Forum for the Future, there has been a profound strategic shift on sustainability; organizations are now including sustainability in their business by building sustainability into their business strategies (Bent and Draper, 2007). Sara Parkin, one of the leading advocates on sustainability leadership, argues that 'leadership is a vital ingredient for achieving sustainability and without which sustainability will never make it in government or business organizations' (2010:89).

It is important to know that 'sustainability leadership' should not be seen as a different school of leadership, but rather a particular combination of individual leadership characteristics applied within a definitive context. This approach is more aligned with the contingency school of leadership theories in terms of approach and characteristics (Courtice, 2011). In addition Ferdig (2007) believes that sustainability leadership can be claimed to involve anyone who seeks sustainable change in an organization regardless of his or her role or position in that organization, and such leaders can connect with others using different assumptions about how people work together to create meaningful change. Even though Redekop (2007) comments that the characteristics of a sustainability leader have not been systematically researched,

Table 1.2 Characteristics of Sustainability leadership

Characteristic of Sustainability leadership	*Source*
• Genuine commitment – sustainability anchored at top management and reflected in core values • Seamless integration - sustainability is aligned with corporate strategy, incorporated in products & services, and embedded throughout operations and processes • Employee engagement - Employees are motivated and participate, sustainability incorporated in individual performance goals and recognition • Transparency and disclosure - Systematic internal and external reporting, documenting success and challenges	Rueegg, 2012
• Mindfulness - staying aware of and paying close attention to the present moment • Advocacy - arguing in favour of sustainability • Femininity - displaying and using characteristically feminine attributes	Ceasar, N. (2011)
• Collaborating and influencing • Change leadership- communicate a compelling vision • Strategic orientation • Commercial orientation-generate value for the organization • Result driven-Translate sustainability vision into a comprehensive programme	Lueneburger and Goleman, 2010
• Acting with integrity • Caring for people • Demonstrating ethical behaviour • Communicating with others • Taking a long-term perspective	Hind *et al.*, 2009
• Having passion and vision. • Systems thinking with a long time horizon. • Innovation and a willingness to learn. • Participatory organizational culture	Hughes and Hosfeld, 2005
• Questioning business as usual • Understanding the role of each player in society and their interaction • Building and managing stakeholder relationships • Respecting diversity • Taking strategic view	Wilson and Holton, 2003

Middlebrooks (2009) have identified the characteristic of sustainability leadership as the ability to see organizational culture through the informed lens of the triple bottom line of sustainability. In addition, the knowledge and understanding of the different balances and interconnections between bottom lines in the pursuit of sustainable ends, the desire to make a positive difference in the long term, the ability to influence in a socially just manner and the ability to manage behavioural and systems change are the characteristics of a sustainability leader (Middlebrooks, Miltenberger, Tweedy, Newman and Follman, 2009). Some of the other characteristics shared by sustainability leaders include innovation and a willingness to learn, having a passion and vision, willingness to teach others, systems thinking with a long time prospect and fostering participatory organizational culture (Hughes and Hosfeld, 2005). A detailed table illustrating characteristics of sustainability leadership is presented in Table 1.2.

1.7 Summary

In conclusion, the process of essential change such as a change towards sustainable built environment begins with a strategic vision that leaders have for their organizations. Leadership is vital in the built environment and a key success factor in the drive towards sustainability. Leadership within built environment organizations has a role in helping to promote sustainable construction by providing training and awareness on sustainability, developing strategies and producing guidance notes and policies. An organizational leadership style could influence the successful implementation and integration of sustainability practices within an organization. Leadership style is also an important part of leadership but there is no one best style of leadership and leaders should be flexible and match their styles with each different situation. Organizational leaders charged with the promotion of sustainable construction practices in the built environment adopt different styles in their desires to embed sustainability practices in their organizations. A strategic leader inspires others to take the appropriate action, with the best interests of the business, the people and the planet in mind whilst transformational leaders create a vision, empower followers and develop a spirit of cooperation and transactional leaders tell others what to do in order to be rewarded and recognize their accomplishments. However, laissez-faire leaders require little of others, are content to let things ride and let others do their own thing. Finally, organizational leadership should fundamentally change the way construction organizations operate from focussing on the short-term maximization of shareholders' returns to paying critical attention to the economic, social and environmental impact of their operations to the society. This will require sustainability leadership with the ability to develop an organizational culture through the informed lens of the triple bottom line of sustainable development.

References

Antonakis, J., Avolio, B. and Sivasubramanium, N. (2003). "Context and leadership: An examination of the nine-factor full-range leadership theory using the Multifactor Leadership Questionnaire", *The Leadership Quarterly*, 14(3), pp. 261–95.

Avery, G. C. (2004). *Understanding leadership: Paradigms and cases*. London: Sage.
Avolio, B. J. (2005). *Leadership development in balance: Made/born*. Mahwah, NJ: Lawrence Erlbaum.
Avolio, B. J. and Gardner, W. L. (2005). "Authentic leadership development: Getting to the root of positive forms of leadership", *The Leadership Quarterly,* 16(3), pp. 315–38.
Avolio, B. J., Luthans, F. and Walumba, F. (2004). "Authentic leadership: Theory-building for veritable sustained performance", Working paper, Gallup Leadership Institute, University of Nebraska, Lincoln.
Bass, B. M. and Avolio, B. J. (1994). *Improving organizational effectiveness through transformational leadership*. Thousand Oaks, CA: Sage.
Bass, B. M. and Avolio, B. J. (1993). "Transformational leadership: A response to critiques", In M. M. Chemers and R. Ayman (eds) *Leadership theory and research: Perspectives and directions*. San Diego, CA: Academic Press.
Bass, B. M., Avolio, B. J, Yung, D. and Berson, Y. (2003). "Predicting unit performance by assessing transformational and transactional leadership", *Journal of Applied Psychology*, 88(2), pp. 207–18.
Bass, B. M. and Riggio, R. E. (2006). *Transformational leadership*, 2nd edn, Mahwah, NJ: Lawrence Erlbaum.
Bass, B. M. and Steidlmeier, P. (1999). "Ethics, character, and authentic transformational leadership behaviour", *The Leadership Quarterly*, 10(2), pp. 181–217.
Bent, D. and Draper, S. (2007). *Leader business strategies: Profitable today, sustainable tomorrow*. London: Forum for the Future.
Bossink, B. A. G. (2007). "Leadership for sustainable innovation", *International Journal of Technology Management and Sustainable Development,* 6 (2), pp. 135–49.
Brown, M. E. and Treviño, L. K. (2006). "Ethical leadership: A review and future directions", *The Leadership Quarterly*, 17(6), pp. 595–616.
Brown, M. E. and Treviño, L. K. (2009). "Leader–follower values congruence: Are socialized charismatic leaders better able to achieve it?" *Journal of Applied Psychology*, 94(2), pp. 478–90.
Bryman, A. (1992). *Charisma and leadership in organizations*. London, Sage.
Bryman, A. (2004) "Qualitative research on leadership: A critical but appreciative review", *The Leadership Quarterly,* 15(6), pp. 729–69.
Bryman, A., Stephens, M. and Campo, C. (1996). "The importance of context: Qualitative research and the study of leadership", *The Leadership Quarterly*, 7(3), pp. 353–70.
Burns, J. M. (1978). *Leadership*. New York: Harper & Row.
Ceasar, N. (2011). *Characterising leadership for sustainable development.* Available at: http://www.guardian.co.uk/sustainable-business/leadership-sustainable-development-characteristics. (Accessed on 20/02/2014.)
Chan, P. W. C. and Cooper, R. (2007). "What makes a leader in construction? An analysis of leaders in the UK construction industry", *Proceedings of the CIB World Building Congress: Construction for development.* 14–18 May 2007.
Courtice, P. (2011). "The challenge to business as usual", *A journey of a thousand miles: The state of sustainability leadership 2011*. Cambridge: University of Cambridge.
Daft, R. L. (2005). *The leadership experience,* 3rd edn, Mason, OH: Thomson South-Western.

Dearlove, D. and Coomber, S. (2005). "A leadership miscellany", *Business Strategy Review,* 16(3), pp. 53–58.
DeChurch, L. S., Hiller, N. J, Murase, T, Doty, D. and Salas, E. (2010). "Leadership across levels: Levels of leaders and their levels of impact", *The Leadership Quarterly,* 21(6), pp. 1069–85.
Dess, G. G. and Picken, J. C. (2000). "Changing roles: Leadership in the 21st century", *Organizational Dynamics,* 29(4), pp. 18–33.
Doh, J. (2002). "Can leadership be taught? Perspectives from management educators", *Academy of Management Learning and Education,* 2(1), pp. 54–67.
Fellows, R., Liu, A. and Fong, C. M. (2003). "Leadership style and power relations in quantity surveying in Hong Kong", *Construction Management and Economics*, 21(8), pp. 809–18.
Ferdig, M. (2007). "Sustainability leadership: Co-creating a sustainable future", *Journal of Change Management*, 7(2), pp. 25–35.
Gardner, W. L. and Avolio, B. J. (1998). "The charismatic relationship: A dramaturgical perspective", *Academy of Management Review*, 23(1), pp. 32–58.
Gattiker, T. F. and Carter, C. R. (2010). "Understanding project champions, ability to gain organizational commitment for environmental projects", *Journal of Operations Management,* 28(1), pp. 72–85.
Giritli, H. and Oraz, G. T. (2004). "Leadership styles: Some evidence from the Turkish construction industry", *Construction Management and Economics*, 22(3), pp. 253–62.
Goetsch, D. L. and Davis, S. B. (2006). *Quality management: Introduction to total quality management for production, processing, and services*, 5th edn, Upper Saddle River, NJ: Pearson Prentice Hall.
Hayibor, S., Agle, B. R., Sears, G.J, Sonnenfeld, J. A. and Ward, A. (2011). "Value congruence and charismatic leadership in CEO–Top Manager relationships: An empirical investigation", *Journal of Business Ethics*, 102(2), pp. 237–54.
Hind, P. *et al* (2009). "Developing leaders for sustainable business", *Corporate Governance*, 9 (1), pp. 7–20.
Hughes, P. and Hosfeld, K. (2005). *The leadership of sustainability: A study of characteristics and experiences of leaders bringing the "triple-bottom line", to business.* Seattle, WA: The Center for Ethical Leadership.
Lehmann, J-P. (2008). "What is responsible business leadership in the early 21st century? Are we heading for a spring of hope or a winter of despair?", Available at: http://www.ind.ch/research/challenges/TC058-08. (Accessed on 15/4/14).
Lloyd-Walker, B. and Walker, D. (2011) "Authentic leadership for 21st century project delivery", *International Journal of Project Management*, 29(4), pp. 383–95.
Lueneburger, C. and Goleman, D. (2010). "The change leadership sustainability demands", *MIT Sloan Management Review,* 51(4), pp. 48–55.
May, D. R., Chan, A. Y. L, Hodges, T. D. and Avolio, B. J. (2003). "Developing the moral component of authentic leadership", *Organizational Dynamics*, 32(3), pp. 247–60.
McCann, J. T. and Holt, R. A. (2010a). "Defining sustainable leadership", *International Journal of Sustainable Strategic Management*, 2(2), pp. 204–10.
Middlebrooks, A., Miltenberger, L., Tweedy, J., Newman, G. and Follman, J. (2009). "Developing a sustainability ethic in leaders", *Journal of Leadership Studies*, 3(2), pp. 31–43.
Munshi, N., *et al.* (2005). "Leading for innovation", *AIM Executive Briefing*, Advanced Institute of Management Research (AIM), London.

Newton, S. (2009). "New directions in leadership", *Construction Innovation*, 9(2), pp. 129–32.

Nicolaou-Smokoviti, L. (2004). "Business leaders' work environment and leadership styles", *Current Sociology*, 52(3), pp. 407–27.

Northouse, P. G. (2010). *Leadership: Theory and practice*, 5th edn. London: SAGE.

Ofori, G. (2009). "Leadership and construction industry development in developing countries (Keynote Presentation)", In A. Rashid (ed).*Proceedings of the CIBW107 International Symposium on construction in developing economies: Commonalities among diversities*, p. 118.

Ofori, G. and Toor, S. R. (2008). "Leadership: a pivotal factor for sustainable development", *Construction Information Quarterly*, 10(2), pp. 67–72.

Oke, A. E. and Gbadura, I. H. (2010). "An examination of project management leadership styles of Nigerian quantity surveyors", *Journal of Building Performance*, 1(1), pp. 57–63.

Opoku, A. (2012). "Promoting sustainability practices through leadership within UK construction organizations", Unpublished doctoral thesis, University of Salford, Manchester-Salford.

Opoku, A and Fortune, C. (2011). "The implementation of sustainable practices through leadership in construction organizations", In C. Egbu and E. C. W. Lou (eds) *Procs 27th Annual ARCOM Conference*,. 1145–54.

Parkin, S. (2010). *The positive deviant: Sustainability leadership in a perverse world.* London: Earthscan.

Quinn, L. and Baltes, J. (2007). *Leadership and the triple bottom line: Bringing sustainability and corporate social responsibility to life.* Greensboro, NC: Center for Creative Leadership.

Redekop, B. (2007) "Leading into a sustainable future: The current challenge", In N. Huber and M. Harvey (ed) *Leadership: Impact, culture, and sustainability*, pp. 134–46. College Park, MD: International Leadership Association.

Riches, C. (1997). "Managing for people and performance", In T. Bush and D. Middlewood (ed) *Managing people in education.* London: Paul Chapman.

Riggio, R. E., Zhu, W., Reina, C. and Maroosis, J. A. (2010). "Virtue-based measurement of ethical leadership: The Leadership Virtues Questionnaire", *Consulting Psychology Journal: Practice and Research*, 62(4), pp. 235–50.

Rueegg, S. (2012). *Global sustainability leaders: How do they do it*? Available at: http://www.britcham.be/LinkClick.aspx?fileticket=Fq727i0pMNA%3D&tabid=192. (Accessed on 10/03/12.)

Sev, A. (2009). "How can the construction industry contribute to sustainable development? A conceptual framework", *Sustainable Development*, 17(3), pp. 161–73.

Tabassi, A. A. and Abu Bakar, A. H. (2010). "Towards assessing the leadership style and quality of transformational leadership: The case of construction firms of Iran", *Journal of Technology Management in China*, 5(3), pp. 245–58.

Taylor, A. (2008). "Promoting sustainable practices: The importance of building leadership capacity", *Proceedings of the Enviro 08 Conference*, Melbourne, Victoria.

Taylor, A.,Cocklin, C., Brown, R. and Wilson-Evered, E. (2011). "An investigation of champion-driven leadership processes", *The Leadership Quarterly*, 22(2), pp. 412–33.

Toor, S. R. and Ofori, G. (2006). "An antecedental model of leadership development", In *Proceedings of joint international symposium of CIB working commissions W55/W65/W86,* Rome, Italy.

Toor, S. R. and Ofori, G. (2008). "Leadership for future construction industry: Agenda for authentic leadership", *International Journal of Project Management,* 26(6), pp. 620–30.

Vecchio, R. P. and Boatwright, K. J. (2002). "Preferences for idealised styles of supervision", *The Leadership Quarterly,* 13(6), pp. 643–71.

Wilson, A. and Holton, V. (2003). "Changing manager mindsets–Report of the Working Group on the Development of Professional Skills for the Practice of Corporate Social Responsibility", The Corporate Responsibility Group, Department of Trade and Industry, London.

Yang, R-L., Huang, C-F and Wu, K-S (2011). "The association among project manager's leadership style, teamwork and project success", *International Journal of Project Management,* 29(3), pp. 258–67.

Yukl, G. (1998). *Leadership in organizations,* 4th edn, Upper Saddle River, NJ: Prentice Hall.

Yukl, G. (2006). *Leadership in organizations.* New York: Elsevier.

Zaccaro, S. J. (1996). "Models and theories of leadership", *US Army Research Institute for the Behavioral and Social Sciences,* Alexandria, VA.

Zhu, W., Riggio, R. E., Avolio, B. J. and Sosik, J. J. (2011). "The effect of leadership on follower moral identity: Does transformational/transactional style make a difference?" *Journal of Leadership and Organizational Studies,* 18(2), pp. 150–63.

2 Behaviourism versus leadership

A transformational need for sustainability in the built environment

Fidelis A. Emuze

2.1 What is sustainability in the built environment?

A consensus definition of 'sustainability in the built environment' appears to be nonexistent (Gadakari *et al.*, 2013), although derivable descriptions are myriad within the literature. This lack of consensus may have arisen from the similar fate of the term 'sustainable development', which also lacks a consensus connotation because of the perceived difficulty in the values that would underlie such an endeavour (Dresner, 2002). Sustainable development, according to the Brundland report (WCED, 1987), represents an approach that meets the needs of the present without endangering the needs of the future. Sustainability with its social, economic and environment strands (Elkington, 2004) which are now joined by resilience and regeneration (Cole and Du Plessis, 2011), is a complex subject that is constantly evolving in the built environment.

However, there are different opinions and definitions for sustainability as each discipline tends to develop its own definition, objectives and agenda to achieve sustainability (Ganah *et al.*, 2008; Gadakari *et al.*, 2013). In essence, sustainability is far too wide an issue that cannot be defined narrowly. While the terms 'sustainability' and 'sustainable development' convey different meanings to different people, at the core of each definition is the concern about how present decisions affect future well-being (Atkinson, 2008). More so, the view that sustainability has different explanations based on ideological alignment calls for principles that must be clearly articulated in order to use the concept of sustainability and permit informed debate from different perspectives. According to Davidson and Venning (2011), a situation that discourages changes to established attitudes and behaviours could persist due to the lack of definitional clarity and theoretical robustness.

The built environment itself represents a multifaceted system that places substantial pressure on the wider environment. As an illustration, buildings and immovable structures have major environmental impact during their life cycle (which includes materials extraction and component manufacture, construction, operation and demolition). A review of construction management literature suggests that sustainable built environment encompasses the creation of buildings and environments that use minimal energy, water in operation, waste in construction and operation and renewable material sources. For such an environment to come into being, actors in the construction industry need to recognise relevant policies, legislation and regulation, as well as understanding current evidence and thinking pertaining to climate change, energy, water, land, pollution, waste, biodiversity, ecology and efficient use of materials within the built environment. This is because the size and nature of building stock has resulted in very high energy consumption, a large carbon footprint and a significant contribution to climate change (Sheth *et al.*, 2008).

Given that sustainability demands a shift to a new perspective within the need to: i) establish the human place in the ecosystem by living in harmony with nature and ii) integrate continuing socioeconomic development with environmental protection (Adetunji *et al.*, 2003), a contextualised definition of sustainability in the built environment is proposed by the author of this chapter. The definition, which states that '*sustainability in the built environment can be defined as an approach to construction, building and development which is embedded in the environmental, economic, social, technological, regenerative, adaptive and resilient initiatives that accommodate the immediate and future needs of a community*' takes cognisance of contemporary literature.

This definition (as illustrated in Figure 2.1) shows a view of sustainability that moves beyond a simplistic model of achieving balance between economy, society and environment to a model based on resilience and adaptive capacity between humans and the natural environment to attain the regeneration of a social–ecological system (Du Plessis and Cole, 2011). The definition also adds technology (Gadakari *et al.*, 2013) as well as regeneration and resilience (Du Plessis and Cole, 2011) to the three traditional pillars of sustainability (Elkington, 2004). This definition includes people, planet and prosperity as equal fundamentals of sustainable development (Roorda *et al.*, 2012).

Improvements in sustainable building design and construction practice can thus be made within the aforesaid norms and conventions by establishing and maintaining a symbiotic and regenerative relationship between human society and the social–ecological systems within which society is embedded (Du Plessis and Cole, 2011). Du Plessis and Cole (2011) further argue that this shift in approach is anchored on a systems view that suggests that effective change will happen through changing the mindset and values of stakeholders, redefining who qualifies as stakeholders and stakeholder roles, and understanding the fact that actors integrated within a construction project team would exhibit different value sets on which to base decisions, and may introduce a cascading set of changes in the rules and goals of the system. The inherent potential of green, sustainability and regenerative design approaches to create the necessary changes in performance improvements in a timely manner has equally been argued (Cole, 2012). Therefore, the roles of stakeholders in engendering sustainability have to change in response to adaptation and regeneration. The contribution of stakeholders to the development and implementation of a sustainable built environment is crucial to deliver meaningful change (Feige *et al.*, 2011).

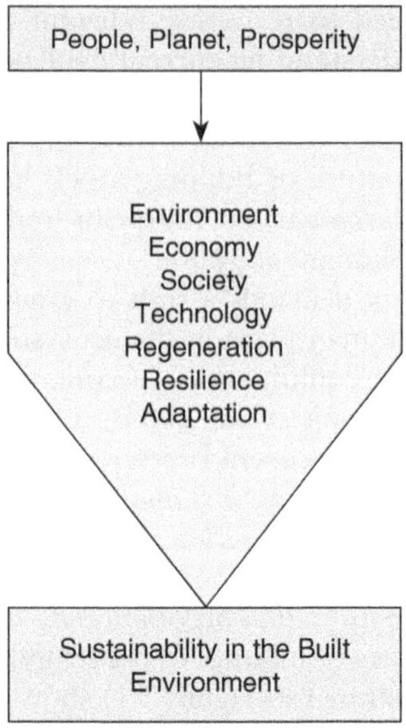

Figure 2.1 Illustrated definition of sustainability in the built environment

The presence of all affected parties can facilitate the development of buildings that are sustainable in the adaptive, social, environmental, ecological, technological and economic dimensions. A change in the behaviour and espoused values of actors is a prerequisite for improved sustainability (Cole, 2011; Nishida and Hua, 2011). The trio of 'value', 'behaviour' and 'leadership' could act collectively to motivate the intended change in paradigm. The central argument in this chapter is that construction leaders do influence organisational effectiveness in the realisation of expected sustainability values. Through flexible leadership style, barriers to sustainable development may be surmounted, especially when the behaviours of project leaders are in tandem with sustainability ethos.

The approach to sustainability that is explained in this chapter is conceptualised at the organisational level in construction management, and it includes four basic variables: i) value ii) behaviour iii) leadership decisions and influence and iv) sustainability. Within the realm of construction management, the organisational level embraces both project and business aspects of construction (Smallwood, 2006). Construction management impacts on sustainability in the built environment through the actions and inactions of project actors (Pryke, 2012). While the construction client is held responsible for a variety of issues in a project (Republic of South Africa, 2003), the project supply chain can either be client-led or contractor-led (Radosavljevic, 2012). The main propositions of this idea are presented in the following sections, along with examples of research that support these propositions and suggest that:

- Espoused values of construction stakeholders determine their commitment to sustainability
- Behaviours of construction stakeholders resonate with the extent that sustainability is promoted within a community
- Contemporary leadership thinking influences the values and behaviours of construction stakeholders in terms of sustainability
- Contextualised barriers to sustainability can be addressed through an appropriate leadership approach.

2.2 Value and sustainability

The concept of value in construction has always had a link with price (Ang, 2008), although some authors have argued that it is relational in nature in terms of satisfying the conflicting needs of a firm and the resources required to meet these needs (Venkataraman and Pinto, 2008). According to Maia *et al.* (2011: 30), '*value is a relation that is established between a subject and an object*'. However, it is important to note that the concept of value has been constrained by the lack of an overarching understanding of what value is, and the predominance of a client-focused perspective of it (Salvatierra-Garrido *et al.*, 2010). Thus, value for construction clients is created through projects and programmes that are undertaken by people working in relationships (Pryke, 2012). Price has always held a high level of consideration among project sponsors and the contractors that deliver construction products. The complex relation between value, price and cost is always demonstrated by treating price as a simplified tuning mechanism between desirable demand and available supply (Ang, 2008). Less resources consumed equates to greater value derivable from a project (Venkataraman and Pinto, 2008). Value can also be considered through the lens of objectivity, subjectivity, relativity, dynamism and setting dependence, while cost, quality and functionality have always been used as measurable attributes of value (Salvatierra-Garrido and Pasquire, 2011). The concept of quality goes together with value for money as perceived by the client (Harris *et al.*, 2013). The aim of construction management is vital to the creation and sustenance of value for clients and other stakeholders. For example, a proposed theory of construction management accepts the fact that construction faces inherent problems that are seemingly difficult to avoid except through informed decisions (Radosavljevic, 2012). Within this theory, value is created within a project (Radosavljevic, 2012) by:

- Reducing the number of construction teams involved
- Improving the quality of the relationships between the construction teams

- Reducing the performance variability between construction teams
- Reducing the external interface experienced by the construction teams.

Furthermore, the concept of value has found expressions and interpretations within sustainability discussions in the construction management research domain. Lean construction researchers have looked at the concept and suggest that there appears to be a synergy between lean and sustainability from a value paradigm perspective. Novak (2012) and Salvatierra-Garrido and Pasquire (2011) explored the synergy between lean and sustainability through the view of the construct of value. These two studies revealed that lean construction practices can significantly contribute to the sustainable development of the built environment. With the use of three exemplary lean projects, Novak (2012) was able to conclude that the construct of value could serve as a catalyst, which shifts construction management from 'restrictive' overtones to a paradigm of sustainable prosperity. The restrictive overtones refer to the tension between cost, time and quality that constrains the delivery of value. Of the three projects that were examined by Novak (2012), project two was noted for 'shared committed leadership'. In the project, the level of stakeholder engagement was similar in terms of project phases, practices, scope, structure and leadership since the stakeholders on the project actively leveraged the synergy between lean and sustainability. In particular, Novak (2012) noted that the project stakeholders understood the link between value from the project perspective and sustainability from the global perspective. Novak (ibid) explored a proposition that links three areas of inquiry as illustrated in Table 2.1. The systemic linkage starts with lean and ends with sustainability, after passing through the construct of value.

The linkage in Table 2.1 indicates that added value is obtainable in the interrelation of product, process and people (Ang, 2008). According to Ang (ibid), added value requires an attitude and culture that go beyond an immediate project as it requires the establishment of shared values, which goes beyond process, money and risks. In other words, it calls for the letting-go of the old paradigm of project value (time, cost and quality) and the acceptance of a

Table 2.1 Logical link between lean, value and sustainability

Notion	*Logical thinking*
Lean	Increased lean cohesiveness can harvest waste and reveal synergies for added project value.
Value	A continuum from project-centred goals to a vision of sustainability.
Sustainability	The vision of sustainability, which brings project values into the right perspective while providing a framework for lean thinking at the same time.

Adapted from Novak (2013: 54)

vision that promotes sustainability (Huovila and Koskela, 1998). The value proposition in the built environment can thus be changed. Based on the illustration of Moffatt and Kohler (2008), environmental effects and flow quantities can be aggregated to assess the performance of entire stocks of buildings and their associated infrastructures. Physical flows for buildings can be valued to ascertain if they satisfy fixed constraints over time. The net performance of the built environment can therefore be measured in terms of combined natural, social and cultural capital, which can be assessed from differentiated value systems (use and non-use values and bequest values) expressed by material and immaterial criteria (Moffatt and Kohler, ibid).

2.3 Behaviourism and sustainability

Within the realm of contingency theory, tasks-oriented behaviour, relations-oriented behaviour, participative leadership behaviour and contingency reward behaviour are often used to explain effective leadership (Yukl, 2011). Early research on effective leadership has shown the implicit assumption that leadership behaviours are positively related to subordinate performance and/or satisfaction in all situations (Yukl, 2011). However, when viewed from a lens that exemplifies a critical perspective, constituted power relations discreetly shape behaviour in organisations (Gordon, 2011). In other words, power is embedded in an organisation's antecedents, meaning systems, socio-cultural norms and discourse that the members of the firm tend to reflect upon in order to make sense of existing work relations and settings (Gordon, ibid). Behaviours of construction stakeholders that assume leadership roles thus have implications for how their followers act within a project environment (Northouse, 2010). Notable authors in psychology have affirmed the potency of behaviourism in terms of conditioning, which is inspired by associations that occur in response to stimuli. For example, the work of Ivan Pavlov, the Nobel laureate, and John Watson shows that behaviour can be a function of the environment in which it is embedded (cf. Theron, 2003). Pro-social behaviour when pursuing capital can be explained by emphasising the role of power in human action (van Aaken *et al.*, 2013). According to van Aaken *et al.* (ibid), pro-social activities can be conceptualised as social practices that individual managers employ in an effort to attain social power through the particular features of the social field – the individual managers' socially shaped dispositions and their stock of different forms of capital. By combining these theoretical concepts, these authors highlight the interplay between the economic and non-economic motivations that underlie corporate social responsibility while acknowledging the perceived influences on deterministic and voluntary aspects of human behaviour.

As a result of these perceptions, it can be said that human behaviour is subject to changes that are dependent on situational context. Construction management researchers contend that behaviours shape attitude toward risks treatment (Qu and Loosemore, 2013), contracts and procurement (Crowe and Fortune, 2012) and innovation and culture (Horsthuis *et al.*, 2012; Ankrah *et al.*, 2008), among other things. It has also been observed that applying social cognitive theory to the construction project context, behaviours of project participants can be predicted based upon internal dispositions of the project participants and the situational context. Therefore, it is possible to engender the sustainability agenda when construction actors with leadership roles engage in pro-social behaviour that is motivated by economic and non-economic motivations. The Theory of Planned Behaviour (TPB) is a valuable social-psychological model that can be used to explain the rationale behind this line of thinking and was used by Ajzen (1985; 1991) to explain the understanding of physical activity behaviour at the level of decisionmaking. The theory argues that the intentions to engage in behaviour are the main determinants of actual behaviour. These intentions are often conceptualised as the motivations to perform behaviour and mediate the influence of three TPB constructs on behaviour (Ajzen, 1991).

Attitude is the first construct, which reflects individuals' personal beliefs about enacting target behaviour, For example, an individual who has a positive attitude toward physical activity is more likely to plan to exercise than a person who believes that such exercise does not have 'value'. The second construct reflects the perceived expectations of precise individuals with respect to the adoption of a specific behaviour. Such a specific behaviour depends on whether the individuals' cultural values promotes the physical activity, whether major parties in the situation perform the physical activity and how these major parties react to the behaviour. The third construct is the perceived behavioural control, which is a reflection of the ease or difficulty of performing the behaviour (Ajzen, 1991).

An example of behavioural change can be found in the work of Bichard (2009), which used employee behaviour, motivation and social psychology to interrogate sustainable behavioural change in three built environment firms. Bichard (*ibid*) observed that the application of motivational methods based on social psychology is effective in raising awareness of sustainability issues. As such, leaders of built environment firms that are interested in improving sustainable performance should consider their leadership style and method of message transmission. Here, the link between behaviour, leadership and sustainability was established. However, the degree of influence that stakeholders possess in a construction context differs. The findings of Doloi (2012) show that stakeholders' influence is consistent with power and interests in a project. Findings emanating from this aforementioned work indicate that project owners and decision makers have the most control over direct economic interests in the project and therefore are ranked in the top of stakeholder's influence list. Employees, end-users and contractors in the construction phase of the project have direct interests in the project – although they have limited power in decision making and therefore are less influential than decision makers.

The onus of the enactment of sustainable behaviour change thus lies with decision makers and project leaders. Apart from project owners, contractors have the leverage to pursue behaviour change among their employees, if they desire to embrace sustainability values. According to Bichard (2009), built environment firms have to believe that employee-driven sustainable change is possible through changing attitudes in the workplace. The argument is anchored on the notion that most people spend more time at work, and there is decent scope for improving attitudes and opinions about the environment and social matters in the workplace (Bichard, ibid). Most importantly, it has been argued that integrated teams are more likely to produce more comprehensive outputs for sustainable projects and that teams with a higher degree of interaction and leadership provide creative sustainable solutions and have better articulation for every aspect of a project (Korkmaz and Singh, 2012).

2.4 Contemporary leadership thinking and sustainability

Contemporary thinking about leadership development views it as a continuous and ongoing process in a lifetime because all experiences have the potential to contribute to learning and progress (Day, 2011). Leadership theories have mentioned individual traits, leadership style, the nature of the task and the environment as the determinants of a leader's effectiveness. For instance, contingency theory describes how aspects of the leadership situation alter a leader's influence on an individual subordinate (Yukl, 2011). Contingency theory suggests that in order to achieve best performance, the leadership style should match situational demands.

The dynamics of the leadership style in construction has made a case for behaviour modification attributes through a manager's supportive, directive, task-oriented and participative roles that influence desired outcomes (Hammuda and Dulaimi, 1997). In other words, the fruitful outcome of any construction works could depend largely on the leaders' ability to address all situations and contingencies that occur. This view resonates with the perception that leadership-based solutions may be more successful in the practice of construction management due to minimal decisionmaking, management and external control risks that manifest (Kashiwagi *et al.*, 2004) and enhanced construction and organisational efficiency (Sullivan *et al.*, 2006). Changing organisational structure and the dynamics of change make leadership by adaptation increasingly appealing and appropriate for the construction industry - especially purposeful leadership with value (Newton, 2008). In answering the question of '*how leaders can influence organisational effectiveness*', Yukl (2008) contends that effective performance requires cooperative efforts, which are flexible and adaptive to a given situational change, by the leaders in a firm. The observation by Yukl (ibid) was reinforced with the argument that a range of leadership behaviours, apart from management systems, can be used to influence expected outcomes.

An example of the potency of contemporary thinking about leadership within the sustainable development world can be illustrated with the work of Dorée *et al.*, (2011). Their work reveals that the rapid adoption of the 'CO_2 *performance ladder*' scheme in the Netherlands has had an impact on improved sustainability within the built environment. The scheme is widely perceived as a meaningful and successful intervention in procurement and tendering; it also serves as an example of successful client leadership in the Netherlands. The study provides valuable lessons about change in the construction industry through the impact of client leadership by showing the strength of certification in conjunction with incentive mechanisms, the importance of institutional embedding, and the need for attention to the support structure for procurement and tendering practices.

2.5 Engendering sustainability in the built environment

Leadership involves a dynamic social interaction in a situational context and both effective followers and leaders are needed (Day, 2011; Northouse, 2010). Leadership is mainly seen as a process of influencing organisational value systems by way of the relationships between leaders and followers so as to attain a predetermined goal (Taylor *et al.*, 2011). The attainment of goals is synonymous with the construction process which entails a dynamic social interaction of human capital and processes in project settings. Thus, in an examination of the dominant methods for conceptualising the built environment, it has been mentioned that it is possible to build models of the environment that assess changes in social, economic and environmental conditions. Such models can be a springboard for understanding the impact of different managerial and social policies at varying levels of interaction (Moffatt and Kohler, 2008). Sustainability of the built environment therefore arguably depends on a fundamental shift in the way resources are used: from non-renewables to renewables, from high levels of waste to high levels of reuse and recycling and from products based on lowest first cost to those based on life-cycle costs and full-cost accounting (especially as applied to waste and emissions from the industrial processes that support construction activity) (Kibert *et al.*, 2000). While the construction literature has shown many barriers to sustainability, the failure to give due consideration to sustainability measures by stakeholders, non-requirement of sustainability measure by clients, failure of stakeholders to enforce sustainable measure and the high cost of sustainability measures, among other barriers (Williams and Dair, 2007) could be ameliorated through behaviour-based interventions. The amelioration should occur with visible features of leadership that provide support to subordinates by modelling expected behaviours, which are highly commended in the management of construction projects (Slattery and Sumner, 2011).

A paradigm shift in the historically restrictive operating mode of managing construction projects to a non-restrictive mode could energise the movement of the construct of value from the 'project view' to a 'sustainability view'. As illustrated in Table 2.2, waste reduction that is based on scarcity should be replaced with resource renewal and optimal use in order to ensure improved sustainability in the built environment context. Hence, instead of limiting the construct of value through prescriptive specifications, construction leaders can make the choice of enhancing value through performance goals. Such goals should embrace an approach to construction and development which is embedded in the environmental, economic, social, technological, regeneration and resilient initiatives that accommodate the immediate and future needs of a community. Table 2.2 was inspired by the notion of delivering value beyond the immovable built environment asset (Novak, 2013). Although clients' requirements will shape different types of construction and contractors may be driven by a particular value proposition (Langford *et al.*, 2003), sustainable development of the built environment can be better impacted when such a proposition is inclusive of sustainability prosperity. Embracing the proposition by construction leaders, at the project and business levels, should foster processes that influence the behaviours of internal and external stakeholders in favour of sustainability (Yukl, 2006).

While it is recognised that firms often deliver value in a variety of ways (Walker, 2012), the construct of value can be mediated by the strategic sustainability stance of clients and contractors. The importance of sustainability should be communicated by construction leaders as they strive to establish organisational behaviours or culture through management decisions that promote sustainability (Avery, 2005). The emphasis on leadership is vital since the success of sustainability will depend on its promotion within government and industry (Parkin, 2010). Through business decisions and

Table 2.2 Transitional mode of construction management

Restrictive mode	*Non-restrictive mode*
Risk aversion exemplified in silo activities.	Collaborative, shared risk / reward, and integration.
Value is constrained by the iron triangle of cost, quality and time.	Value is viewed beyond the iron triangle of cost, quality and time.
Value is limited by prescriptive specifications.	Value is enhanced by performance goals.
Waste reduction is championed by scarcity.	Waste reduction is about resource optimal usage and renewal.
Minimisation of carbon footprint of operational activities.	Economic, environmental, social, technology, regeneration and adaptive balance.
Reductionist benchmarks in the form of 'Green Buildings'.	A systems thinking or perspective of sustainability in the built environment.

Inspired by Novak (2013: 52)

organisational activities, construction leaders should embed sustainability in the fabric of their internal and external commitments (Ofori and Toor, 2008). For example, the work of Quinn and Dalton (2009) supports the perception that leaders within firms that formally promote sustainability practices have had to integrate social, economic and environmental consideration into the visions, values and operations of their firms to obtain desired outcomes.

2.6 Concluding remarks

To remain competitive in an unsettled and uncertain environment, the construction industry requires leaders who are flexible and adaptive as situations arise. Construction leaders should be able to understand the complex relationships among project imperatives that generate contextualised value in the process and to acknowledge the behaviours and leadership styles that would best fit situations that would not marginalise sustainability. By clarifying how construction leaders can influence and improve sustainability, the central argument in this chapter offers a new way of thinking in construction management. The discussion has pointed to how flexible leadership can promote a more sustainable built environment. In particular, sustainability requires construction leaders to reconsider, reform and restructure their projects and organisations. The construct of value should be revisited to reflect the expectations of sustainability. This construct is client and contractor-led and should cascade down to all involved in a construction project life cycle.

Motivating change that allows sustainability to flourish in the built environment requires clarity of how value is conceptualised. The processes of change also rely on the behaviour of leaders – hence, the need for construction leaders to embrace the concept and advance new ways of thinking in the built environment. In the construction process, activities and decisions need to seek to overcome barriers to sustainability through teamwork that is led by leaders with flexible and adaptive traits. Through flexible leadership (that influences the behaviours of project teams), project stakeholders should be able to move from the restrictive to the non-restrictive mode of managing the business and construction projects. When implementing a business strategy that considers the triple aspect of people, planet and profit, construction leaders should be able to align the diverse motivation of stakeholders to a value proposition that promotes sustainability. This approach would weave sustainability into the fabric of each firm involved in a construction venture. In other words, in the face of snowballing competition for human capital and other resources, decisions that are embedded within a system that recognises the new construct of value have the propensity to better inform a sustainable future in the built environment.

Note

The author wishes to express gratitude to Professor David J. Edwards, Professor and Head of Faculty Research at the Birmingham City Business School, and Dr Torben V. Rasmussen, Senior Researcher at the Danish Building Research Institute, for their very helpful comments and suggestions on earlier version of this chapter.

References

Adetunji, I., Price, A., Fleming, P. and Kemp, P. 2003. "The application of systems thinking to the concept of sustainability", In Greenwood, D. J. (ed) *Procs 19th Annual ARCOM Conference*, 3–5 September 2003, University of Brighton, Association of Researchers in Construction Management, Vol. 1, pp. 161–70.

Ang, G. 2008. "Competing revaluing construction paradigms in practice", P. Barrett (ed), *Revaluing construction*. Oxford: Blackwell Publishing, pp. 83–104.

Ankrah, N. A., Proverbs, D. G. and Ahadzie, D. K. 2008. "Exploring the behaviours of construction project participants through social cognitive theory", In: Dainty, A. (ed) *Procs 24th Annual ARCOM Conference*, 1–3 September 2008, Cardiff, UK. Association of Researchers in Construction Management, Vol.1, pp. 443–53.

Ajzen, I. 1985. "From intentions to actions: A theory of planned behavior", In Kuhl, J. and Beckmann, J. (eds) *Action-control: From Cognition to Behavior*, Heidelberg: Springer, pp.11–39.

Ajzen, I. 1991. "The theory of planned behavior", *Organizational Behavior and Human Decision Processes,* 50(2), pp. 179–211.

Atkinson, G. 2008. "Sustainability, the capital approach and the built environment", *Building Research and Information*, 36(3), pp. 241–47.

Avery, G. 2005. *Leadership for sustainable futures: Achieving success in a competitive world*, Cheltenham, UK: Edward Elgar.

Bichard, E. 2009. "The application of sustainable behaviour change strategies in three built environment companies", *Journal of Engineering, Design and Technology*, 7(1), pp. 7–20.

Cole, R. J. 2011. "Motivating stakeholders to deliver environmental change", *Building Research and Information*, 39(5), pp. 431–35.

Cole, R. J. 2012. "Transitioning from green to regenerative design", *Building Research and Information*, 40(1), pp. 39–53.

Crowe, P. and Fortune, C. 2012. "A preliminary method of classifying collaborative contractual behaviour in higher education construction projects", In Smith, S. D (ed) *Procs 28th Annual ARCOM Conference*, 3–5 September 2012, Edinburgh, UK, Association of Researchers in Construction Management, pp. 901–11.

Day, D. V. 2011. "Leadership development", In Bryman, A., Collinson, D. L., Grint, K., Jackson, B. and. Uhl-Bien, M. (eds) *The Sage handbook of leadership*, London: Sage, pp. 37–48.

Davidson, K. M. and Venning, J. 2011. "Sustainability decision-making frameworks and the application of systems thinking: an urban context", *Local Environment*, 16(3), pp. 213–28.

Doloi, H. 2012. "Assessing stakeholders' influence on social performance of infrastructure projects", *Facilities*, 30(11/12), pp. 531–50.

Dorée, A., van der Wal, G. and Boes, H. 2011. "Client leadership in sustainability: How the Dutch railway agency created CO2 awareness in the industry", In Egbu, C. and Lou, E.C. W. (eds) *Procs 27th Annual ARCOM Conference*, 5–7 September 2011, Bristol, UK, Association of Researchers in Construction Management, pp. 685–94.

Dresner, S. 2002. *The principles of sustainability*. London: Earthscan.

Du Plessis, C. & Cole, R. J. 2011. "Motivating change: shifting the paradigm", *Building Research and Information,* 39(5), pp. 436–49.

Elkington, J. 2004. "Enter the triple bottom line", In Henriques, A. and Richardson, J. (eds) *The triple bottom line, does it all add up? Assessing the sustainability of business and CSR*. London: Earthscan, pp. 1–16.

Feige, A., Wallbaum, H. and Krank, S. 2011. "Harnessing stakeholder motivation: Towards a Swiss sustainable building sector", *Building Research and Information,* 39(5), pp. 504–17.

Gadakari, T., Mushatat, S. and Newman, R. 2013. "Intelligent buildings: key to achieving total sustainability in the built environment", *Journal of Engineering, Project, and Production Management*, 4(1), pp. 2–16.

Ganah, A., Pye, A. and Hall, G. 2008. "The role of knowledge transfer in sustainability research in the built environment discipline", In Dainty, A. (ed) *Procs 24th Annual ARCOM Conference,* 1–3 September 2008, Cardiff, UK, Association of Researchers in Construction Management, pp. 299–307.

Gordon, R. 2011. "Leadership and power", In Bryman, A., Collinson, D. L., Grint, K., Jackson, B. and Uhl-Bien, M. (eds) *The Sage handbook of leadership*, London: Sage, pp. 195–202.

Hammuda, I. M. and Dulaimi, M. F. 1997. "The effects of the situational variables on the leadership styles in construction projects", In Stephenson, P. (ed) *13th Annual ARCOM Conference*, 15–17 September 1997, King's College, Cambridge, Association of Researchers in Construction Management, Vol. 1, pp. 22–31.

Harris, F. and McCaffer, R. with Edum-Fotwe, F. 2013. *Modern Construction Management*. 7th edition. Chichester, UK: Wiley-Blackwell.

Horsthuis, C., Thomson, D. & Fernie, S. 2012. "The case for slack to promote innovative behaviour in construction firms", In Smith, S. D. (ed) *Procs 28th Annual ARCOM Conference*, 3–5 September 2012, Edinburgh, UK, Association of Researchers in Construction Management, pp. 533–42.

Huovila, P. and Koskela, L. 1998. "Contribution of the principles of lean construction to meet the challenges of sustainable development", In *Proceedings of the 6th Conference of the International Group for Lean Construction* (IGLC), 13–15 August, Guaruja, Brazil.

Kashiwagi, D., Egbu, C., Kovel, J. and Badger, W. W. 2004. "Leadership vs. management in the construction industry", In Khosrowshahi, F (ed) *20th Annual ARCOM Conference*, 1–3 September 2004, Heriot Watt University, Association of Researchers in Construction Management, Vol. 2, pp. 1005–15.

Kibert, C. J., Sendzimir, J. and Guy, B. 2000. "Construction ecology and metabolism: natural system analogues for a sustainable built environment", *Construction Management and Economics*, 18(8), pp. 903–16.

Kormaz, S. and Singh, A. 2012. "Impact of team characteristics in learning sustainable built environment practices", *Journal of Professional Issues in Engineering Education and Practice*, 138(4), pp. 289–95.

Langford, D., Martinez, V. and Bititci, U. S. 2003. "Best value in construction", In J. Kelly, Morledge, R. and Wilkinson, S. (eds) *Best value in construction*. Oxford: Blackwell, pp. 1-11.

Maia, S., Lima, M. and Neto, J. P. B. 2011. "A systemic approach to the concept of value and its effects on lean construction", In *Proceedings of the 19th Conference of the International Group for Lean Construction* (IGLC), 13–15 July, 2011, Lima, Peru, pp. 23–32.

Moffatt, S. and Kohler, N. 2008. "Conceptualising the built environment as a social-ecological system", *Building Research and Information*, 36(3), pp. 248–268.

Newton, S. 2008. "Changing the framework for leadership in the construction industry", In Dainty, A. (ed) *Procs 24th Annual ARCOM Conference*, 1–3 September 2008, Cardiff, UK, Association of Researchers in Construction Management, pp. 433-442.

Nishida, Y. and Hua, Y. 2011. "Motivating stakeholders to deliver change: Tokyo's cap-and-trade program", *Building Research and Information*, 39(5), pp. 518–533.

Northouse, P. G. 2010. *Leadership: Theory and practice*, 5th edition, London: Sage.

Novak, V. M. 2012. "Value paradigm: Revealing synergy between lean and sustainability", In: *Proceedings of the 20th Conference of the International Group for Lean Construction* (IGLC), 18–20 July, 2012, San Diego, California, USA, pp. 51–60.

Ofori, G. & Toor, S. R. 2008. "Leadership: A pivotal factor for sustainable development", *Construction Information Quarterly*, 10(2), pp. 67–72.

Parkin, S. 2010. *The positive deviant: Sustainability leadership in a perverse world*. London: Earthscan.

Pryke, S. 2012. *Social network analysis in construction*. Chichester: Wiley-Blackwell.

Qu, Y. and Loosemore, M. 2013. "A meta-analysis of opportunistic behaviour in public-private partnerships: manifestations and antecedents", In Smith, S. D and Ahiaga-Dagbui, D. D (eds) *Procs 29th Annual ARCOM Conference*, 2–4 September 2013, Reading, UK, Association of Researchers in Construction Management, pp. 415–24.

Quinn, L. and Dalton, M. 2009. "Leading for sustainability: Implementing the task of leadership", *Corporate Governance*, 9(1), pp. 21–38.

Radosavljevic, M. 2012. *Construction management strategies: A theory of construction management*, Chichester: Wiley-Blackwell.

Republic of South Africa 2003. Government Gazette No 25207: Construction Regulations 2003. Pretoria.

Salvatierra-Garrido, J., Pasquire, C. and Thorpe, T. 2010. "Critical review of the concept of value in lean construction theory", In *Proceedings of the 18th Conference of the International Group for Lean Construction* (IGLC), 13–15 July, 2010, Haifa, Israel, pp. 33–41.

Salvatierra-Garrido, J. and Pasquire, C. 2011. "The first and last value model: sustainability as a first value delivery of lean construction practice", In *Proceedings of the 19th Conference of the International Group for Lean Construction* (IGLC), 13–15 July, Lima, Peru, pp. 1–11.

Sheth, A., Price, A. D. F., Glass, J. and Achour, N. 2008. "Reviewing the sustainability of existing healthcare facilities", In Dainty, A (ed) *Procs 24th Annual ARCOM Conference,* 1–3 September 2008, Cardiff, UK, Association of Researchers in Construction Management, pp. 1193–1202.

Slattery, D. K. and Sumner, M. R. 2011. "Leadership characteristics of rising stars in construction project management", *International Journal of Construction Education and Research*, 7(3), pp. 159–74.

Smallwood, J. J. 2006. "The practice of construction management", *Acta Structilia*, 13(2), pp. 62–89.

Sullivan, K. T., *et al.* 2006. "Leadership, the information environment, and the performance measuring project manager", In Boyd, D (ed) *Procs 22nd Annual ARCOM Conference*, 4–6 September 2006, Birmingham, UK, Association of Researchers in Construction Management, pp. 113–22.

Taylor, A., Cocklin, C., Brown, R. and Wilson-Evered, E. 2011. "An investigation of champion-driven leadership processes", *The Leadership Quarterly*, 22(2), pp. 412–33.

Theron, A. 2011. "Perspectives on general and work behaviour", In Bergh, C. Z. and Theron, A. L. (eds) *Psychology in the work context*. New York: Oxford, pp. 3–15.

van Aaken, D., Splitter, V. and Seidl, D. 2013. "Why do corporate actors engage in pro-social behaviour? A Bourdieusian perspective on corporate social responsibility", *Organisation*, 20(3), pp. 349–71.

Venkataraman, R. R. & Pinto, J. K. 2008. *Cost and value management in projects*, New Jersey: John Wiley & Sons.

Walker, D. H. T. 2012. "Innovation and value delivery through supply chain management", In Akintoye, A., Goulding, J. and Zawardie, G. (eds) *Construction innovation and process improvement*. Chichester: Wiley-Blackwell, pp. 125–53.

WCED (1987) *Our common future: report of the World Commission on Environment and Development*. New York, United Nations General Assembly.

Williams, K. and Dair, C. 2007. "What is stopping sustainable building in England? Barriers experienced by stakeholders in delivering sustainable developments", *Sustainable Development*, 15(3), pp. 135–147.

Yukl, G. 2006. *Leadership in organizations*. New York: Elsevier.

Yukl, G. 2008. "How leaders influence organisational effectives", *The Leadership Quarterly*, 19(6), pp. 708–22.

Yukl, G. 2011. "Contingency theories of effective leadership", In Bryman, A., Collinson, D. L., Grint, K., Jackson, B. and Uhl-Bien, M. (eds) *The Sage handbook of leadership*, London: Sage, pp. 286–98.

3 Leadership: a negation of agency

Ciaran McAleenan and Philip McAleenan

3.1 Introduction

The full development of sustainability in the built environment, during construction and post construction stages requires that leadership, as a catalyst, is relinquished in favour of the collective agency of the mature reasoning group, whether that group is the design team, the onsite works teams or the wider partnership of stakeholders as exists at the relevant stages of design, construction and use (McAleenan and McAleenan, 2009b). The title of leadership is valued as it often confers authority on the decisions, pronouncements and actions of those regarded as leaders. With it comes respectability and an expectation that those not in the leadership position will defer to the authority of the leading opinion. In some instances the cloak of leadership provides and is actively used as a protective barrier against objections and opposition.

The general concept of leadership is multifaceted and commonly has many uses and meanings (Eacott, 2013 and 2014); it may be applied to those who are decision makers or to those holding office who acquire the role of a leader by dint of said office. In some cases leadership is assumed or assigned to those with ownership of or control over large-scale industrial or financial enterprises. Intellectual or thought leadership is claimed by others, including institutions, some of which may, on the basis that they have more members or have greater recognition than comparable bodies, set objectives and lay claim to being thought leaders in particular fields of expertise. Yet leadership has the potential to negate agency, and this is crucial to understanding the limited role that it should have in human affairs. Using a critical analysis approach, this chapter investigates the leadership concepts prevailing in the built environment today, arguing that the imposition of a 'leader' negates autonomy. It explores how treating leadership as a function, which transfers to those most suited to exercise it, presents the opportunity to develop autonomous actions for sustainable construction and development.

3.2 Aspects of leadership

The professional and academic discourse on leadership is substantial and encompasses a range of conflicting theoretical perspectives (Spoelstra, 2013;

Eacott, 2014), from how it is to be defined to whether leaders are born or made (Mostovicz *et al.*, 2009), from where it is to be found (Kumar, 2012) to what role it has in modern society (Chari, 2012; Lowder, 2014). A brief examination of a few of the different uses to which the term leadership is applied illustrates the complexity of the issue.

Transactional leadership is based on the idea that there exists a formal authority within which the goals and objectives and the conditions that the members of the organisation must agree to and follow in order to gain reward for their effort are set by those at the top of the organisation. It includes sanctions, generally one-way in employer/employee relationships, for those who would breach the agreements or otherwise contravene often-unwritten rules about what is expected of them.

Much of the task of transactional leaders is to bring order to the organisation and improve the efficiency of the organisation through the development of rules and procedures. Efficiency-oriented organisations tend towards a bureaucratic system as paper work is generated to detail the rules and procedures and to prove that they have been followed, or to prove that someone else was at fault if they were not followed. Recent history in the UK's approach to health and safety at work resulted in an overabundance of paperwork to prove that risk assessments had been carried out and there was evidence that this was the result of over interpretation of the regulations (UK Parliament Works and Pensions Committee, 2008). This was in all likelihood a case of efficiency leading to inefficiency, and it is a problem that stems from the focus on efficiency and process rather than on effectiveness; once the goal has been established, the leader, determined to preserve his authority as leader, maintains his commitment to the goal and his authority through discipline and steadfastness.

An effective leader is focussed on the objectives, and the process evolves naturally and efficiency arises through the efforts of followers whose rights to have and interpret their own experiences are respected, or at least not denied (Mostovicz *et al.*, 2009). This approach fits into *the leader-member exchange (LMX) theory* that views the relationships as complementary albeit hierarchical. It is an interpersonal relationship between the superior and the subordinate partners and is one based on quality; it is more than a transactional relationship where one does what is required in exchange for an agreed remuneration. Here the exchange is concerned with the nature of the relationship with positive support being provided to the subordinate, including respect and trust. This environment in promoting effectiveness advances the success of the organisation and builds loyalty and commitment.

The strength in the LMX approach lies in the quality of application and the extent to which it is applied in the organisation. Research has shown that high-quality relationships reap benefits for the organisation in the form of increased commitment from employees who will go beyond the requirements of their contract (Ilies *et al.*, 2007). Conversely, poor-quality application of LMX will have a reduced effectiveness as it will fail to promote the positivity necessary for good relationships, and where it is applied to select in-groups within the organisation the out-groups will notice and feel disconnected, leading to a sense of being less or considered less than the in-group. This in turn will lead to practices, including unethical practices, which are not in accord with the organisational objectives and will shift the leadership into a transactional mode in order to tackle the disciplinary matters that occur.

In some assessments transformational leadership is seen to be akin to LMX in that it involves quality relationships, but in the view of Spoelstra (2013) it departs from the LMX approach in that the relationships go beyond exchange; leaders and followers, he states, '*overcome exchange and self-interest in a manner that resembles Marion's giver or givee: their subjectivity is bracketed, so that they open themselves up to givenness*'[1]. Despite what he views as entailing a pseudo-religious conversion, the practice of transformational leadership is concerned with the identification of qualities within the subordinate, the follower, that indicate the potential for leadership and the subsequent cultivation of those qualities with the intention of developing future leaders. It is considered a positive form of leadership in that it meets the organisational performance objectives while at the same time safeguarding the organisation from ethical scandals. It is in this perspective an ethics-based leadership in that it is concerned with interests beyond the organisation and how it fits into or at least is viewed by wider societal stakeholders.

Objectively though, *transformational leadership* is concerned with moulding the follower and potential future leader in its own image. The organisation and its good standing in the public mind take precedence over the individual experiences and worldview of the follower. Codes such as the OECD Principles of Corporate Governance (2004) and the UK Code of Corporate Governance (FRC 2012) recognise that the financial scandals of the past 15 years are unacceptable, but rather than questioning the objectives of the businesses that were central to those scandals, the codes lay down guides for business and financial leaders to continue what they were doing but with a more ethical approach that takes into consideration the requirements for social responsibility. Transformational leadership does not question the objectives of the organisation, but seeks qualities in the subordinates that will favour the ethical achievement of them.

The leadership theories described, whether transactional, LMX or transformational in nature, are all top-down approaches requiring subordinates to defer to the final decision of the supervisor or manager. There are gradations on the degree of veracity in this assessment, depending upon the industry. Thus, for example in high hazard/low risk industries such as nuclear and nuclear new build, the level of competency required is extremely high and coming with it is the requirement that all members of the construction and industrial teams feed into the defence in depth and safety programmes to a greater extent than would be found in lower hazard industries and construction projects (Petrangelli, 2006). Work on the Finland's Olkiluoto 3 plant, which began in 2005 and is projected to be completed by 2009, has experienced many complaints over design flaws, poor quality control and construction lapses. Writing in the *Guardian* newspaper (Kirby, 2014), University of Oxford researcher Peter Wynn Kirby explained that the problems included poorly trained subcontractors who have poured substandard concrete and welded containment vessels improperly. The automatic control system and safety system were found to be insufficiently robust and the company was sent back to the drawing board. Similar problems and similar vast cost over-runs are experienced at the Flamanville site in France.

With nuclear-generated power considered one of the arms of a sustainable energy future, it is not unreasonable to expect that the construction processes would adhere to the strict defence in depth principles necessary to avert the type of disasters that have already been experienced at Three Mile Island, Chernobyl and Fukushima, or that with the number of nuclear plants worldwide, the cost overruns, which are to be expected, would be kept to a reasonable sum rather than almost three times the original budgets. What we see here is a failure in leadership, both political and organisational, to design, develop and deliver on commitments to the public, in the process submitting current and future generations to the fear of a catastrophic failure with the consequent severe human and environmental damage.

3.3 Political vs. organisational leadership

In the sphere of the built environment, the decision making capacity of communities may be compromised by the overriding decisions of 'political-leaders' based on their commitments to other agents rather than to matters of generating sustainable communities. Flyvbjerg and Stewart (2012) indicated that no city that has hosted the Olympics since 1960 has done so without massive cost overruns, and many cities and nations have learnt to their detriment how risky the financial decisions of such mega-projects have been. London 2012 was on track to becoming the most costly Olympics ever and with the most significant cost overruns since 1999, was reversing a positive trend in falling cost overruns for the Games.

3.3.1 Political leadership

In the political sphere the claim of leadership is made on the basis of politicians having more votes than their opponents and/or being voted into office by the electorate. In a survey of business managers prior to the 2010 general election in the UK, the Institute of Leadership and Management (ILM) concluded that 'leadership matters, particularly in politics, where it builds consensus in the party, balances competing agendas and ultimately wins elections' (ILM, 2010). This latter point is, however, a conflation of electoral politics and assumptions about the voting intentions of the electorate with the electorate's desire or otherwise to be led by those they vote for. The ILM identified 5 dimensions to leadership in order to determine what the Leadership Quotient would be of party 'leaders'. Their five dimensions of leadership were identified as:

1 General ability – overall competence and capability in role, together with skills and knowledge to perform to a high standard;
2 Integrity – the importance of being honest, principled and fair. If management is about doing things right, leadership is about doing the right thing, setting an example and telling people the truth;
3 Vision – fundamental to leadership, this is about seeing the big picture, identifying opportunities and driving followers to achieve. Successful leaders consult before making decisions and explain what they are doing and why;
4 Communication – how well they present and absorb information. Communication is a two way street – good leaders ask for opinions and listen. When they have decided what to do, they explain clearly in language people understand;
5 Engagement and commitment – how well leaders translate their vision into a clear sense of purpose in their individual followers, generating personal commitment to make it happen. (For a short critique of the Five Dimensions, see Box 1)

To determine where UK and international political and business leaders fell, 2,000 UK practicing managers were asked to rate, on a scale from 0 to 10, a number of well known so-called leaders on all of these five dimensions. The five ratings were averaged together to create a single Leadership Quotient (LQ). On the basis of these five dimensions the three main UK party leaders scored rather low compared to other world leaders such as Barack Obama and Angela Merkel, with business leader Richard Branson scoring highest of all.

Unsurprisingly, each UK leader was scored highly by those whose voting intentions were towards their party of choice (ILM, 2010), but the report does not state whether the voting intentions were determined on the basis of prior or post evaluation of the party leaders. The failing of this approach to leadership determination is that it is based on the subjective perceptions of one particular demographic.

Box 3.1 Critique of the Five Dimensions of Leadership

An assessment of the Five Dimensions of Leadership shows that they are not unique to leadership and can be found in many areas and activities of life. Competent people in whatever field of endeavor must demonstrate optimal skills and capability, up to the point where they are competent to become competent — that is, they have achieved such a high level of capability that they are able to become more than what they were by their own efforts.

(Continued)

(Continued)

Integrity is a reasonable expectation of every adult that does not require leaders to demonstrate by example how to be truthful. In 2009 it could have been questioned whether integrity is regarded as a quality necessary for political leadership, given the scandals then surrounding parliamentary expenses and the efforts taken in the House to prevent the disclosure of parliamentarians' expenses under the Freedom of Information Act.

Communication is a two-way dialogue but the meaning given in this dimension to the two-way street is a self-contradiction; there is one way listening and one way telling. Consultation is about seeking the views of other before making decisions, but this is not leadership in the sense that everyone is going voluntarily towards the same objective. This approach has a greater or lesser degree of compulsion in that political decisions taken at government level do not and cannot have absolute consensus. The notion that successful leaders consult before making decisions is negated in reality. Political leaders rarely consult and yet can be seen as successful. Indeed, some are seen as successful because of the very absence of consultation; they have a vision, they take a risk and if it pans out, they are "successful". In politics success is often a matter of how constituencies view the outcome of political decisions, a matter of "one man's meat . . . "

Having these dimensions as a person does not in themselves make a leader. The dimensions may contribute to the quality of a leader, but that then raises the question as to whether it is possible to have a bad quality leader.

Referring to prime ministers and presidents as world leaders is based as much if not more on their decision making roles on behalf of their respective countries as it is on their personal share of the vote and their popularity. This conflates their role with the notion that their fellow nationals will follow or agree with their decisions or where the nation is being led. Within the ILM model, consultation is an integral element of being a visionary and in that context it is not unreasonable to expect world leaders to hear and accept the opinion of experts in, for example, the field of climate change and to arrive at a vision for a sustainable future complete with agreed programme for achieving such. However, commenting on the Copenhagen climate conference in 2009, the economist Joseph Stiglitz (2010) noted the failure of the world leaders to achieve a number of essential agreements on how to achieve the goal of saving the planet, on carbon reductions and on the fair allocation of sharing the burden of cost. World leadership in this context turns out to be not so much about leading the electorate into a safe and sustainable future, but about protecting the interests of the industrial and financial sectors of the respective countries. '*Even the commitment of the accord to provide amounts approaching $30 billion for the period 2010-12 for adaptation and mitigation appears paltry next to the hundreds of billions of dollars that have been doled out to the banks in the bailouts of 2008–09*' (Stiglitz, 2010).

3.3.2 Industry leadership

The conflation of decision making roles or positions with leadership is also found in industry where CEOs and senior management are similarly regarded as leaders and in the recruitment process qualities that make up leadership are central to decisions on whether to appoint or not (Fresh Minds, 2010a), particularly in the appointment to management and executive positions. Leadership qualities are seldom a requirement for technical and trade positions, suggesting an inherent bias in leadership theory that leadership cannot generally come from the 'lower ranks'. An exception is that in the case of the military services forms of leadership are trained in all ranks and qualities of leadership is sought for and rewarded through promotions from all ranks.

3.3.3 Power relationships

It is the case that decisions by politicians and CEOs take the country/organisation in particular directions but this is not necessarily leadership. Such decision makers have the power and authority to take the country/organisation along particular routes regardless of the support that they may have or not for the decision. The power relationship between 'leader' and 'follower' is not often central to the discourse but is evident within the terminology of differing theories; thus, whether the leader, as in employer/manager, controls the follower/employee through a basic leader-member-exchange (LMX) relationship (e.g. work carried out in exchange for a wage) (Tummers and Knies, 2013) or in a transformational leadership approach within which the leader supports and encourages the growth of subordinates to, possibly, in turn become leaders/managers in the company/organisation (Latour and Rast, 2004) or at least to enjoy their position and be supportive of the company (Reid), the relationship is always one of power; the 'followers' are not equal to or fully equal to the 'leader'. These power relationships are expanded in the managerial structures established by organisations and the role or function that individuals are assigned in those structures.

In industry, and this is as true for the construction industry as any other, the position of 'leader' and 'follower' parallels the relationships that people have to the means of production; that is, there are those who own and exercise control over the capital assets necessary for production, including the raw materials and sources of raw materials, the factories, plant and equipment, and the finance capital necessary to produce wealth, and those whose employment enables production to take place. Simply put, those who own the company have the authority to decide how it is to be run and who does what, and this is fundamentally an LMX relationship at its least sophisticated, though in practice it has evolved much more on the back of management and leadership theory and through the engagement of professional services to facilitate effectiveness.

3.3.4 'Safety' leadership

Managers and supervisors are expected to have leadership qualities in order to ensure that the workforce of the 21st century is well managed and that the companies continue to compete successfully in the market (ILM, 2012; Conchie and Moon, 2010). For example, an aspect of this is the role that health and safety plays in the success of a company and thus its sustainability. Some of the safety discourse centres on safety leadership and the role of the safety leader, including who is to be regarded as the safety leader. Legislation in the UK mandates that companies with boards appoint a member of the board to have responsibility for ensuring that safety is incorporated into board reports and discussions – thus in the context of the above, ensuring that the leaders of the company are themselves the leaders of safety. Based on this the Institution of Civil Engineers (Northern Ireland Region) Expert Panel on Health and Safety initiated a series of events in 2012 that brought the leaders of the construction industry together with academic and professional bodies to explore new ways of improving safety in the industry. The initiative commenced with the signing of a declaration by these leaders to commit themselves and their organisations and companies to the improvement of safety and to lead others in the industry to follow suit (McAleenan and McAleenan, 2012). Continuing with the safety example, the Institution of Occupational Safety and Health (IOSH) views transformational leadership as being more effective in achieving accident prevention in the reduction in unsafe behaviours by employees because it engenders trust and respect from employees, citing a positive relationship between supervisors' safety leadership styles and employees' safety behaviour (IOSH, 2010).

In the former example, the function of the leader is to lead by example. The CEOs of the largest construction companies had direct influence on their own business and how it was to be conducted, but also, by dint of size and the fact that they worked with or engaged many other smaller companies in the industry, they were in a position to influence a substantial portion of the industry and this influence was deemed more effective when they themselves set the example. In the latter example, without negating the role of the board and chief executive in leading the company, IOSH views the leadership function as being devolved throughout and exercised in a supportive and encouraging manner.

3.4 Leadership and governance

Nonetheless, success and sustainability is not simply a case of how companies manage their workforces, but also how they manage their business overall. In the early 2000s in the financial sector, corruption led to the downfall or bankruptcy of some of the world's largest corporations, once thought too big to fail (Enron, Arthur Andersen and WorldCom). Following on from this the Organisation for Economic Cooperation and Development (OECD) developed the Principles of Corporate Governance designed to ensure good practice by corporations in their financial dealings. Governments followed suit with the introduction of appropriate legislation including the UK's Corporate Governance Code. This code includes principles of leadership that require the sharing of leadership responsibilities so that no one individual has unfettered decision making powers, and principles regarding effectiveness and accountability to company shareholders for the success of the company, but, unlike the OECD principles, the UK code does not require consultation with stakeholders beyond shareholders (OECD, 2004, FRC 2012).

Despite these efforts, the world recession commencing in 2007 resulted from a crisis in a number of financial and investment practices, some of which were very risky, that resulted in increased foreclosures in mortgaged properties, the increase in personal debt, the withdrawal of financial support to investors and developers and the resultant collapse or near-bankruptcy of major financial institutions and banks. This led to detrimental effects on industry, including construction, and with substantial government bailouts of banks, there was little public finance left to support investment in infrastructure and other public works projects, normally seen as critical to the economy. At the same time organisations such at the International Labour Organisation (ILO) recognised that the recession, particularly if was to be extended, would lead to a decrease in good health and safety practice resulting in more workplace fatalities and injuries and warned of this at the 2008 World Congress on Safety and Health at Work in Seoul, South Korea.

3.4.1 Ethical leadership

Emerging perspectives on how companies should behave challenge unrefined approaches to directing and managing organisations. More frequently in the early 21st century, ethical leadership is being promulgated as central to what a leader does (Kumar, 2012; Chari, 2014; May and Pardey, 2013). Though ethics is substantially explored in the theological and philosophical literature stretching back millennia, in the modern world when business leaders substitute values for economic performance, the general welfare of society decreases (Chari, 2014). Thus ethical leadership theory developed out of the failures of the corporations in the early 2000s and the codes of governance that were being advocated by national and international bodies.

Particular focuses were on ethical practices and sustainability, taking their cue from requirements of industry leaders to consider the rights of and consult with stakeholders (OECD, 2004). Analysing the functions of businesses three objectives can be identified that all companies must consider to be sustainable, namely, that they exist for the good of the company, the good of the employee and the good of society: the Triple Objectives. The good of society as a function of ethical business was recognised by Boardman and Lyon (2006) when they stated that any business acts with the consent of society and from this it follows that business leaders must have a moral duty not to abuse societal trust.

The function of morality/ethics is to guide human behaviour in a manner that is not simply non-injurious to others but in a way that it actively contributes to the well-being of others (Fromm, 1947). This is not based on a premise that 'what is good for me is good for others', but rather the reverse; it is formed on a knowledge and an understanding of what is good for others in order to determine what is good for the individual. Further than this, morality, as a necessary aspect of what it is to be human, is continuous and thus moral/ethical leadership is, also, necessarily continuous, and it is in this continuity that we have the core of sustainability.

The construction industry globally has a reputation that falls well below what may be deemed ethical, since it is an environment where competition between contractors leads to low price mentalities, fierce competition and paper-thin margins which in turn lead to quality reductions to cut costs and save time (Abdul-Rahman *et al.*, 2007, 2010). Unethical practices at the top by the leadership have the added effect of encouraging unethical behaviour or practices further down the line, though this does not necessarily follow. The ILO's warning at Seoul in 2008 that as the recession hit, good practices in the workplace would suffer as companies struggled to maintain the bottom line raises questions about leadership for sustainability as companies' focus is on individual company success rather than broader societal need.

3.4.2 Thought leadership

Positions of authority based on knowledge and expertise confer leadership status on some, even in situations where the individual is not proactively developing a leadership role or function. This happens when the idea or knowledge that an individual has gains wider support from others who take up those thoughts with the intention of putting them into practice or in some instances developing them further. There is no doubt that when an idea is published there is some element of wanting others to take it up but without an active programme of subordinating followers to the originators' ongoing thoughts on the matter. Leadership here is defined primarily by the nature and actions of the followers.

The expert as a leader, active or passive, is not a new concept, but Thought Leadership is. Leaders Direct (2014) defines it as being 'radically different from traditional top-down leadership. It can be directed up as well as down or sideways, has nothing to do with position or managing people, is the basis of innovative change and is egalitarian because it can shift rapidly from one person to another'. In this respect they state that it is not something that can be monopolised yet it is none the less an aspiration or an objective that some organisations strive of towards and in the process it becomes commodified; IOSH, for example, views the provision of high quality guidance as a key part of the organisation's thought leadership and corporate social responsibility activities.[2]

3.5 Managerialism

Much of what is considered leadership, particularly in industry and commerce, can in reality be described as managerialism. Eacott (2014) described 'leadership' as '*a product of the managerialist project and therefore rather than providing any possible alternative or meaningful way of knowing, doing and being in the social world, "leadership", is merely an extension of managerialism*'.

In the various theories outlined the veracity of this can be clearly seen. In company terms, the leadership manifested at the CEO and the Board levels is in effect charged by shareholders to develop the policies that will ensure the continued success of the company. That they develop good policies and achieve increased market share and profit is not an indicator of good leadership, but rather an indicator of competence.

The senior management team, and this descriptor says it all, develops the practical processes by which the company objectives and policies are to be achieved. That they can do that with the cooperation of managers, supervisors and other employees is again less about leadership competence than it is about management competence coupled with the ability to maintain positive relationships throughout the workforce.

Management is concerned with perpetual improvement, in doing better by pursuing more effective or efficient practices (Fresh Minds, 2010b, Eacott, 2014) and the various forms of leadership up to and including ethical leadership are centred on the pursuit of improved performance. The masses of followers, of employees are in relationships of practical necessity with the industry leaders. The objectives of the company are of less immediate concern that the smooth and effective running of the company and the guarantee of wages in return for effort.

3.5.1 The necessity of agency

The idea that the concept of leadership means in some instances to guide and direct, in others cases to be to the fore or to be the best in the field and in still others to be in advance of others who willingly follow has merit, but limited in that for much of the time what is seen as leadership is in effect aspects of managerialism: that is, people in senior positions in organisations are charged with an objective that requires they organise others in an appropriate manner to achieve those objectives. It does not necessarily require that those who carry out the tasks essential to achieve the objective are in agreement with the objective or are supporters of those senior to them, i.e. the necessary leader/follower dynamic is not met.

Fundamentally, when the concept of leadership is analysed at its core, there lies the notion, first, that leadership involves an idea or a person with a 'vision' that others are willing to adopt or follow. There is a common agreement on the end as well as a broadly shared notion that one or more should take the lead and guide the remainder, whether that person is the originator of the idea or those most capable or organising the followers.

Second, the willingness to follow must be voluntary in nature. This voluntary nature of what could be defined as followership allows for individuals to choose another objective or to change their minds or ways of thinking and opt out; it allows the individual to make the choice of whether or not to follow the leader. Compulsion is a negation of leadership and many of the examples that we have examined are relationships of authority and subordination, where one or more persons have the role of instructing others in what to do and how to do it, and those others do so whether in exchange for wages, for fear of penalty or for some other gain not associated with agreement on the end.

The third component of the analysis is that of the surrender of the function of leadership when the above conditions no longer apply. When an objective is achieved or no longer deemed valid, the function of the leader in that project ends, as do the functions of the followers. When the leader fails to manage the project or loses the confidence of the followers, again the role ends, or passes to another. When the followers decide that they can achieve the objective without any identifiable leader, that they can do it themselves, then the function of leadership is no longer necessary. Leadership can be lost when followers cease to follow for whatever reason, and a critical quality of leadership is recognising when the function of leadership is no longer required. This latter point is often absent from the leadership discourse, though transformational leadership recognises one function of the leader is as a cultivator of successors or those who would be partner leaders (Latour and Rast, 2004; Reid); however, this does not account for conditions when leadership is not required. It is a necessary adjunct to the first two components and should the function of leader remain when one or both of them are removed, then leadership transforms into hierarchical authority and into one of the number of forms discussed above.

Without these elements leadership as generally perceived has the potential to negate agency, whether of the individual or the team. Agency is central to competence inasmuch as the competent person or the competent body is constrained in the ability to take decisions within their particular sphere of competence and influence, and thus becomes dependent upon others for the effective outworking of their activities (McAleenan and McAleenan, 2009a, 2010).

Dependency is a negation of or a limitation on competence and in the sphere of sustainability the decision making capacity of competent people may be compromised if they override decisions of others who may be on the periphery of or outside of particular design and construction activities, or by those who lay claim to greater degree of authority by dint of investment or political position.

Other factors influence the decision making capacity of individuals and organisations. Overtly these may include budgetary limitations or resource allocations set by the finance department, competency levels established by Human Resources or general and specific training deficiencies unmet by the training department. Subliminally, workplace culture will be influenced to a greater or lesser degree by other messages put out by the company such as drops in profitability, fears of redundancies and so on. These messages work to confound compliance instructions leading to a diminution of their decision making capacity potentially putting workers in a position where they may be held responsible for workplace failures in which they have participated (McAleenan and McAleenan, 2013).

Leadership requires an all-party acceptance of and agreement to the achievement of an objective and recognition that within the parties all are due equal consideration and respect as human beings (Kohlberg, 1971). Without this recognition, leadership is either a form of managerialism or a transactional relationship based on authority and wages. Eckensberger (2007) in his work on morality and culture postulates all humans as agents capable of self-reflective action. Recognising that heteronomous decision making arises out of necessities, the developing individual moves from heteronomy to autonomy and in the competent person this development has been achieved (within a particular sphere). In this perspective the competent individual or team is not led but supported and appropriately resourced by the structures within the organisation, albeit their activities are towards ends set by others. In safety, by definition it is the competent worker/team that is the expert and by extension the safety 'leader'. Once set to work, the team collectively assesses the requirements, including the safety requirements to achieve a successful outcome, and collectively sets about achieving that outcome. Team leaders may be established but in this context they act as coordinators or facilitators and are not a negation of agency within the team.

3.5.2 The reasoning person

The discourse on leadership encompasses a range of topics from leadership styles to qualitative leadership, and from the forms of leadership to the qualities that go to make up a leader. Absent from the discourse are matters concerning the nature of human beings and what it is that defines the mature reasoning person. This omission leaves a substantial gap in the discourse, paving the way for an almost absolute focus on the individuals and their capacity to guide, direct or otherwise take groups of others in particular directions. Notwithstanding this, there is some discourse on the group, on followership, and as such is from the perspective of what makes a good follower, i.e. someone who facilitates and eases the tasks the leader in his/her role. Followership is the corollary of leadership, as well as being the obverse of the coin in which at a functional and psychological level a follower accepts the authority of the identified leader and behaves in a manner that makes possible the function of leader, and facilitates him/her in meeting their objectives.

The form and function of both leader and follower that is now explored is in the context of what it is to be a human being and questions whether the current discourse is adequate to the task of creating a sustainable built environment wherein all stakeholders are respected and the hierarchy of human needs is met. The factors that contribute to identifying the mature reasoning person have regard to culture, competence, ethics reasoning and agency.

Culture is at its most fundamental the mode of human existence. It is the product of human action on the world and in turn each person is the product of culture. It is differentiated from nature in that it entails conscious action to transform the environment to meet human needs, a transformation that is the expression of a person's conscious relationship to the world and in which he or she organises himself or herself to confront the challenges of existence, chooses the best response and acts upon it, changing both the world he or she inhabits and himself or herself in the process (Freire, 1973).

Fundamental to culture is morality, the embodiment of self-control, and ethics reasoning, a conscious process of determining what one should or should not do. Geertz (1973) described culture as a system of uniquely human controls. Culture, however, is not just uniquely human; it is unique to each person as each one consciously and individually experiences and interprets culture for himself or herself and makes his or her choices on how to respond. A human [3]therefore exists as a moral being, responsible for and in control of the judgements he or she makes. In order to become what Kant and Locke have described as a self-reflective agent" (Körner, 1995, citing Kant; Katsafanas, 2011), each person has the capacity to act independently on his or her judgements; agency requires that we have autonomy of decision making and action.[3]

There are principles of ethics reasoning, the highest stage of which requires that the mature reasoning person will give equal consideration to all (Kohlberg, 1971). Decision making stems from self-chosen ethical principles that appeal to logical comprehensiveness, universality, and consistency. These principles are not the concrete moral rules as may be laid down by theology but are abstract and ethical and in the sense of Kant's categorical imperative, we do what is right because it is right, respecting the equality and dignity of all human beings as individuals. Thus the mature reasoning agent is one who has attained the level of competence necessary to reflect upon his or her world and to decide the manner in which he or she will respond to it. Collectively, the mature reasoning group will do no less. Acting at the highest level of ethical reasoning, it will behave in accordance with the principle that the best interests of the groups are achieved by acting in the best interests of all: the Fromm (1947) counsel.

In the traditional sense, the relationship between leader and follower negates the concept of agency and by extension is a negation of human capacity in both leader and follower. Allowing a voluntary commitment on the part of the follower, the leader nonetheless relinquishes his or her authority to the leader in respect of going towards a defined goal or objective, even in situations where they are in agreement with that goal. The mature reasoning person, individually and in the collective, will not relinquish the reflective decision making capacity without stepping away from his or her responsibilities to determine what the right objectives are and the right way to achieve them. When he or she does so, his or her reasoning is at a conventional level where he or she acts out of a sense of duty, respecting authority and maintaining social order for its own sake (Kohlberg, 1971). It is a rule based behaviour that is followed whether the rule is right or wrong.

Conversely, the leader, even with the best intentions in the world, is often in the position of setting the goals and determining the means by which they are achieved, e.g. industry leaders and senior executives. As followers, by the nature of followership, are prevailed upon to go by the rules, so in turn does the leader establish what those rules are and in setting the rules must lead by example and follow those same rules. We need but look to the political leaders who are charged with establishing the laws of the land and determining the penalties for non-compliances, and religious leaders who set the moral standards for the faithful and in their theological councils establish the penances for transgressors, to see that rule-making is a function of 'leadership' and in this respect the leader too functions at the same lower levels of reasoning.

In attempting to establish a necessary relationship between leadership and followership it can be discerned that necessity is uni-directional, namely that leadership requires that there are followers, without which the leader is but a lonely figure walking a road of his or her own making. Conversely, followership can exist without the person of a leader. It is enough that individually and collectively reasoning persons can adopt and follow an idea or a vision and collectively decide upon how they will attain their goals without the intervention of a leader. As such followership must not be constrained by compulsion, other than the compulsion that emerges from individual rational decision making at the highest level of ethics reasoning and is imposed on the self by the self.

3.6 Conclusion

Maslow stated, 'What a man can be, he must be'. There is no doubt that the ecological basis of current thinking on sustainability remains an integral element of both environmental and human wellbeing. However the built environment is much more than an interloper on the natural environment; it is the human response to basic and mid-level needs outlined by Maslow for warmth, shelter, security and social contact. Sustaining the built environment reconciles the natural environment's resilience with the economic demands of society and social justice in the present and into the future. The discourse on sustainability in the built environment must thus be directed towards determining what is essential for realising universal principles of justice and reciprocity and respect of the dignity of all as individual persons.

Sustainability is a concept that has validity in itself and not as a policy decision. It affects all equally and without exception. The development of the idea and the achievement of progress towards a sustainable future require the collective agency of society in total, which means recognising that society has the capacity to act as a mature reasoning body. Leadership that acts in an authoritarian manner negates agency and that type of leadership must be relinquished in favour of one that embraces people and causes that affect society. It must be capable of promoting the universal principles described and ultimately it must recognise that its function in human affairs is limited and is exercised at the grace and forbearance of society. Authentic leadership will emerge from the societal capacity for self-determination, devolving managerial functions from time to time to a 'leadership' that is temporary and constrained by the needs and desires of those it serves.

Notes

1 Jean-Luc Marion is a phenomenologist whose central idea is that 'there are phenomena of such overwhelming givenness that intentional acts aimed at them are over-run – saturated'.

2 IOSH. Membership Advisory Panels,http://www.iosh.co.uk/About%20us/Be%20part%20of%20IOSH/Volunteer%20opportunities/Member%20advisory%20panels accessed 28 March 2014

3 However from a Nietzschean perspective it is argued that our actions are causally determined by factors other than conscious choice, that we act in the way we do out of the 'iron hands of necessity', influenced by our feelings, and motives. The will, it seems, is not as free as we may think (Katsafanas, 2014).

References

Abdul-Rahman, H. *et al.*, (2007), "Does professional ethic affect construction quality?", *Quantity Surveying International Conference*, 4–5 September 2007, Kuala Lumpur, Malaysia.

Abdul-Rahman, H., Chen, W. and Xiang, W. Y. (2010), "How professional ethics impact construction quality: Perception and evidence in a fast developing economy", *Scientific Research and Essays,* 5(23), 3742–49.

Boardman, J. and Lyon, A. (2006), "Defining best practice in corporate occupational health and safety governance", HSE UK.

Chari, V. (2012), "21st century moral leadership: Why resignation has become a strategic option for corporate leaders", [Online], available at: https://www.academia.edu/5013383/Moral_Leadership_in_the_21st_Century (Accessed on 28 March 2014).

Conchie, S. and Moon, S. (2010), "Promoting active safety leadership", IOSH, UK, 2010.

Eacott, S. (2013), "Leadership' and the social: Time, space and the epistemic", *International Journal of Educational Management*, 27(1), 91–101.

Eacott, S. (2014), "Beyond the hype of 'leadership'", *Perspectives on Educational Leadership, Australian Council for Educational Leaders*, 20(1), 1–3.

Eckensberger, L. H. (2007), "Morality from a cultural perspective", In Zheng, G., K. Leung & J. Adair (eds.), *Perspectives and progress in contemporary cross-cultural psychology*. Beijing: China Light Industry Press, 25–34.

Financial Reporting Council (2012), "The UK Corporate Governance Code", FRC, London.

Flyvbjerg, B. and Stewart, A. (2012). "Olympic proportions: Cost and cost overrun at the Olympics 1960–2012", Saïd Business School, University of Oxford, 2012.

Freire, P. (1976), *Education: The practice of freedom*. London: Writers and Readers Publishing Cooperative.

FreshMinds Research (2010a), *Creating future leaders*. Institute of Leadership and Management (ILM), UK, 2010.

FreshMinds Research (2010b), *Leading change in the public sector*. Institute of Leadership and Management (ILM), UK, 2010.

Fromm, E. (1947), *Man for himself*. London: Routledge Classics. [Published 2003]

Geertz, C. (1973), "Thick description: Toward an interpretive theory of culture", In *The Interpretation of Cultures*. New York: Basic Books.

Ilies, R., Nahrgang, J. D., and Morgeson, F. P. (2007), "Leader–member exchange and citizenship behaviors: A meta-analysis", *American Psychological Association Journal of Applied Psychology*, 92(1), 269–77.

Institute of Leadership and Management (ILM) (2010), *Politics: Leadership matters*. Institute of Leadership and Management, (ILM) UK, 2010.

Institute of Leadership and Management (ILM) (2011), *Index of leadership trust*. Institute of Leadership and Management (ILM), UK.

Institute of Leadership and Management (ILM) (2012), "The leadership and management talent pipeline", Institute of Leadership and Management (ILM), UK.

Kant, I. (1785), *Metaphysics of morals*. Cited in Russell, B. (1947), *A history of Western philosophy*, George Allen and Unwin,

Katsafanas, P. (2011), "Activity and passivity in reflective agency", *Oxford studies in metaethics*, Vol. 6, R. Shafer-Landau (ed.), Oxford: Oxford University Press, 219–54.

Katsafanas, P. (2014), "Nietzsche and Kant on the will: Two models of reflective agency", Forthcoming in *Philosophy and Phenomenological Research*.

Kirby, P. W. (2014), "Europe's new nuclear experience casts a shadow over Hinkley", *Guardian*, 25 March 2014, [Online], available at: http://www.theguardian.com/environment/2014/mar/25/europes-new-nuclear-experience-casts-a-shadow-over-hinkley (Accessed on 26 March 2014).

Kohlberg, L. (1971), "Stages of moral development: The stages of moral development according to Kohlberg", Penn State University, [Online], available at: http://info.psu.edu.sa/psu/maths/Stages%20of%20Moral%20Development%20According%20to%20Kohlberg.pdf (Accessed on 25 March 2014).

Körner, S. (1955), *Kant*. London: Penguin.

Kumar, D. (2012), "Ethical leadership for the 21st century", Amity Global Business School, Chandigarh, IND, November 2012.

Latour, S. M. and Rast, V. J. (2004), "Dynamic followership, the prerequisite for effective leadership", *Air and Space Power Journal*, [Online], available at: http://govleaders.org/dynamic_followership.htm.

Leadersdirect (2014), "Thought leadership", [Online], available at: http://www.leadersdirect.com/thought-leadership (Accessed on 17 January 2014).

Lowder, T. (2007), "Implementing a dynamic leadership program: A moral construct for adding cultural value", [Online], available at: https://www.academia.edu/948795/Implementing_a_Dynamic_Leadership_Program_A_Moral_Construct_for_Adding_Cultural_Value (Accessed on 28 March 2014).

Maslow, A. H. (1943), "A theory of human motivation", Originally published in *Psychological Review*, 50, 370–96.

May, T. and Pardey, D. (2013), "Added values: The importance of ethical leadership", Institute of Leadership and Management (ILM), UK 2013.

McAleenan, P. and McAleenan, C. (2009a), "An exploration of structured and flexible approaches to recognising engineering competence", *Proceedings of CIB W099 Conference*, Melbourne, 2009.

McAleenan, P. and McAleenan, C. (2009b), "Development of the competent company in the context of the Seoul Declaration", Canadian Society of Safety Engineers PDC proceedings, Canada, 2009.

McAleenan, P. and McAleenan, C. (2010), "Calculating your flight distance – the evolution of safety in the competent company", Canadian Society of Safety Engineers PDC proceedings, Canada, 2010.

McAleenan, P. and McAleenan, C. (2012), "Decision Makers Conference, Implementation Report 1", *Institution of Civil Engineers*, NI Region, 2012.

McAleenan, P. and McAleenan, C. (2013), "Maturing workplace culture in the context of evolved ethical agency", Proceedings of CIB World Congress, Brisbane, AU, 2013.

Mostovicz, E. I., Kakabadse, N. K. and Kakabadse, A. P. (2009), "A dynamic theory of leadership development", *Leadership & Organization Development Journal*, 30(6), pp. 563–76.

Organisation for Economic Cooperation and Development (OECD) (2004), OECD Principles of Corporate Governance.

Petrangeli, G. (2006), *Nuclear Safety*. Oxford, UK: Butterworth-Heinemann,

Reid, M. "A critique of Transformational Leadership theory", [Online], available at: https://www.academia.edu/300040/A_critique_of_Transformational_Leadership_theory (Accessed on 28 March 2014).

Spoelstra, S. (2013), "Is leadership a visible phenomenon? On the (im)possibility of studying leadership", *Int. J. Management Concepts and Philosophy*, 7(3/4), 174–88.

Stiglitz, J. E. (2010), "Overcoming the Copenhagen failure", [Online], available at: http://zcomm.org/znetarticle/overcoming-the-copenhagen-failure-by-joseph-stiglitz/#, (Accessed on 25 March 2014).

Tummers, L. G. and Knies, E. (2013). "Leadership and meaningful work in the public sector", *Public Administration Review*, 73(6), 859–68.

UK Parliament Works and Pensions Committee (2008), "Evidence on health and safety", February 2008. [Online], available at: http://www.publications.parliament.uk/pa/cm200708/cmselect/cmworpen/uc246–iii/uc24602.htm, (Accessed on 20 March 2014).

Part II

Sustainable built environment

4 Sustainable development in the UK construction industry

Alex Opoku

4.1 Introduction

The concept of sustainability has been growing in importance and has become the basis of most socio–economic activities and developments in the built environment (Edum–Fotwe and Price, 2009). Sustainability is important for the construction industry because of the impact of construction activities and products on the environment. The pursuit of sustainable construction practices provides numerous opportunities for organizations prepared to take on the sustainability challenge. Construction plays a significant role in the ambition towards sustainable development and should therefore balance its social benefits, economic costs and environmental impacts (Yates, 2003) and the construction industry will be judged on how its activities contribute to the world's sustainable development agenda (Kibert, 2007). Construction organizations can contribute to the sustainable development agenda by adopting sustainable practices in the delivery of construction projects. It is therefore recommended that to achieve sustainable construction project delivery, construction organizations should double their efforts and actively promote sustainability practices throughout the whole project life cycle (Opoku, 2012). Construction organizations should attach equal importance to all sustainability practices in order to fully address the three dimensions of environmental, economic and social sustainability.

4.2 Sustainable development and the construction industry

Sustainable development involves balancing and integrating the economic, social and environmental considerations in all business decisions to enable humanity to satisfy their basic needs and enjoy a better quality of life, without compromising the quality of life of future generations (DTI, 2006). The construction industry significantly impacts on the environment and society (Myers, 2005; Manoliadis *et al.*, 2006). It has a major role to play towards the achievement of sustainable development because it affects water, resources, land use, greenhouse gas emissions (Pitt *et al.*, 2009) as well as communities, and the health of the general public (Sev, 2009; Holton *et al.*, 2008). The construction industry therefore needs to play its required role in the fight towards achieving a sustainable society now and for the future generation.

4.2.1 Sustainable development

Sustainable development has been defined in many ways. Parkin (2000) pointed out that there are well over 200 rumored definitions of sustainable development in circulation. Even though Riedy (2003) believed that the meaning of sustainable development is still strongly contested, the most commonly accepted definition for sustainable development is the one in Brundtland's commission report which defines sustainable development as:

> *"Meeting the needs of the present without compromising the ability of future generations to meet their own needs . . . A process of change in which the exploitation of resources, the direction of investments, the orientation of technological development, and institutional change are all in harmony and enhance both current and future potential to meet human needs and aspirations"* (Brundtland 1987:43).

However, Opoku and Ahmed (2013:141) provide an alternative definition that addresses the concept of need and human behaviour as:

> *"The adjustment of human behaviour to address the needs of the present, without compromising the ability of future generations to meet their own needs".*

Brundtland's (1987) definition aims to be more comprehensive and addresses the key concept of needs while Opoku and Ahmed (2013) put emphasis on human behaviour in an attempt to meets their needs. The transition towards achieving sustainable development requires changes in human behaviour, values and attitudes that will meet human needs. Opoku and Ahmed (2013) further argue that the construction industry needs to establish a common sustainable construction framework that clearly defines sustainable development within the context of construction project delivery. The benefits of committing to sustainability can be enormous and include money saving by reducing waste and increasing efficiencies, risk mitigation, winning more customers or clients and attracting and retaining talented graduates (Miller, 2010).

4.2.2 Sustainable construction

The pursuit of sustainable construction practices provides numerous opportunities for organizations that are prepared to take on the sustainability challenge. There have been several initiatives by the UK government to encourage reform in the construction industry through a number of reports and policies (Latham, 1994; Egan, 1998; DETR, 2000; Fairclough, 2002;

ODPM, 2003). For example in 2000, the UK government introduced a construction-specific strategy, *Building a Better Quality of Life,* which highlighted key themes for action by the construction industry, namely:

- Design for minimum waste;
- Re–use built assets;
- Aim for lean construction and minimize waste;
- Minimize energy in use;
- Preserve and enhance biodiversity;
- Do not pollute;
- Conserve water resources;
- Respect people and their local environment;
- Monitor and report (DETR, 2000).

Sustainable construction is the application of sustainable development principles in the construction industry. There is no universal agreed definition of sustainable construction because it is continuously developing as the concept of sustainability is understood more clearly (UNDESA, 2010). Parkin (2000), describes sustainable construction as a construction process that incorporates the basic themes of sustainable development. Sustainable construction aims at reducing the environmental impact of a building over its entire lifespan, providing safety and comfort to its occupants and at the same time enhancing its economic viability (Addis and Talbot, 2001). The Marrakech Task Force on Sustainable Buildings and Construction set up by the UN Department of Economic and Social Affairs (UNDESA) defines sustainable construction as:

> *"The construction that brings about the required performance with the least unfavourable ecological impacts while encouraging economic, social and cultural improvement at a local, regional and global leve"* (UNDESA, 2010:3).

Carter and Fortune (2003) used a grounded theory approach to explore the perception of sustainable development held by those involved in the procurement process. They argue that understanding what is meant by sustainability in the construction industry, at all levels of project delivery, is key to its successful implementation. Sustainability is about improving economic growth (economy), social progress (equity) and environmental protection (ecology) concurrently. This can be achieved through: energy efficient buildings that have little or no harm on our environment; reducing the consumption of resources; increasing the use of recycled materials; and generally reducing CO2 emissions from buildings (Williams and Sutrisna, 2010). Sustainable construction project delivery can be aided by having a clear sustainability goal at the design stage, sensitive site selection, recyclable materials, the right construction methods in terms of energy and resource efficiency (Singh, 2007). Reffat (2004) argues that the concept of sustainable construction goes beyond environmental issues to include economic and social sustainability issues with the view of adding value to the quality of life of individuals and communities. The sustainable construction dimensions are discussed in the subsequent paragraphs below:

Environmental sustainability addresses the impact of construction activities on the environment by minimizing waste, using natural resources and energy efficiently. In addition, the environmental dimension of sustainable construction involves efficient use of energy and resources that prevents pollution of the environment (Sourani and Sohail, 2011).

The **economic dimension of sustainability** refers to the implementation of construction practices that provide for positive economic growth (Beheiry *et al*., 2006; Jones *et al*., 2010), through job creation, competitive advantage, and reduction in operating and maintenance costs (Baloi, 2003). Sourani and Sohail (2011) describe economic sustainability as focusing on issues such as whole life costing, support of local economies and financial affordability for intended beneficiaries.

Social sustainability deals with legal, moral and ethical obligations of construction organizations to their stakeholders. Sustainable construction considers respect for employees and working collaboratively as well as issues surrounding the working of construction organizations in the local community. Sourani and Sohail (2011) add that social sustainability involves issues such as health and safety, involvement of stakeholders, equality and diversity in the workplace and creating employment opportunities. The key sustainable construction themes and associated issues are presented in Figure 4.1. Construction organizations are now inculcating sustainability practices in the construction process at both pre–construction and post–construction stages of construction project delivery. These sustainable construction practices are discussed in the next section.

4.3 Sustainability practices in construction

Construction organizations committed to the sustainability agenda should always aim at minimizing the impact of their construction activities on the society. In order to minimize the environmental, social and economic impact of construction projects, organizations should adopt sustainable policies and practices in all aspects of design, choice of materials and construction methods. Many construction organizations are now producing building designs that use minimum energy and water resources, produce minimum waste and prevent pollution as well as preserving and enhancing the local ecological biodiversity. Such sustainable practices will therefore minimize the overall negative impact of the built asset throughout its whole life.

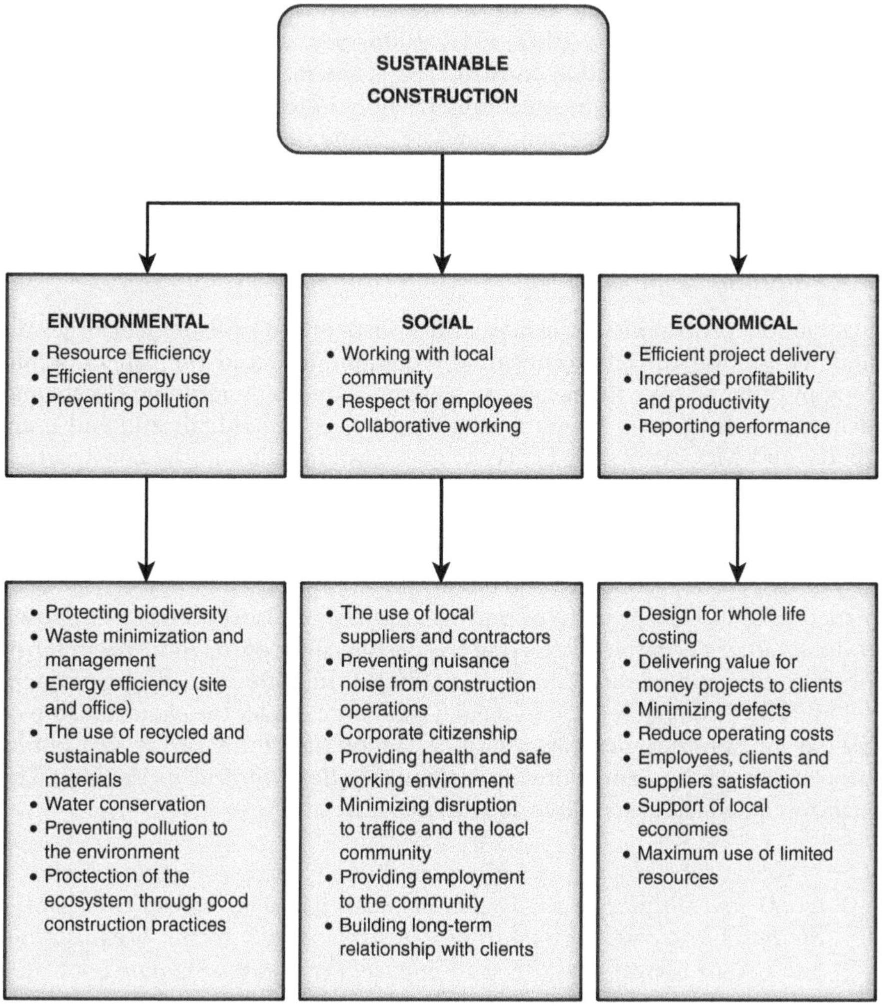

Figure 4.1 Key sustainable construction themes and issues

(Source: Opoku, 2012)

Singh (2007) argues that the promotion of sustainability in the construction industry is all about adopting appropriate practices in terms of our choice of construction materials, the origin of materials used in the industry, methods used for construction and more importantly the design values that minimize waste. The production, transport and use of construction materials have negative impacts on the environment affecting the quality of air, water and soil. Construction operations also use large quantities of energy and generate high levels of noise, waste and dust. This environmental damage is caused by the construction industry over the whole course of a project; therefore, it is essential that sustainability practices are adopted throughout the whole life of the project, from its planning to its demolition phase (Rohracher, 2001; DTI, 2006; Sev, 2009; Son *et al.*, 2011). A key point for sustainable construction is the production of high quality buildings, which are environmentally friendly in terms of energy and water efficiency and the management of waste. Some of the major sustainability practices being promoted in delivery of construction projects are discussed in the subsequent sections.

4.3.1 Sustainable procurement

Sustainable procurement considers the effects of the procurement of goods and services on the environment, the communities and the general economy in order to take the necessary steps to reduce any negative effects and deliver true long-term benefits to organizations, individuals and end users (Berry and McCarthy, 2011). To achieve sustainability in the construction industry, the industry and its stakeholders need to adopt a more holistic approach throughout the supply chain process (Hamid and Kamar, 2012). The integration of sustainability into construction procurement can produce significant social, economic and environmental benefits. However, the construction industry's practice of putting emphasis on lowest price as opposed to best value is negatively affecting the performance of the industry in terms of time, cost and quality (Shafii *et al.*, 2006). In 2006, the UK government set up 'The Sustainable Procurement Task Force' under the chairmanship of Sir Neville Simms to develop a national action plan for ensuring that public procurement fully contributes to sustainable development in the UK. The task force defined sustainable procurement as the:

> *"Process whereby organisations meet their needs for goods, services, works and utilities in a way that achieves value for money on a whole life basis in terms of generating benefits not only to the organisation, but also to society and the economy, whilst minimising damage to the environment"*, (DEFRA, 2006:10).

To successfully implement sustainable procurement practices, it is important that construction organizations understand procurement as a significant integrated process of design, planning, project management, project services, commissioning and all the processes that deliver a complete construction project.

4.3.2 Sustainable design

The adoption of sustainable design and construction is gathering considerable impetus in the construction industry and both the design team and clients are appreciating the increasing benefits of sustainable design in terms of minimal energy usage and decreased negative impact on the environment

(Pulaski, 2004). A sustainable building design that integrates a passive heating and cooling system with sufficient daylight will reduce the cost of energy use to occupants because such design uses less electrical cooling, heating and lighting. Even though the initial cost of a sustainably designed construction may be high, money can be saved in terms of energy bills over the whole life of the building, paying for the use of systems that are more efficient in the long run (LDSA, 2008). Sustainable design only considers the current economic viability of a design option but also measures its long–term economic and environmental impacts through the adoption of a whole life–cycle approach to design (Fenner *et al.*, 2006). Sustainable design goes beyond just the design of the building to include the choice of construction material specifications, as well as engineering systems and equipment used. A sustainable design must aim to eradicate the negative impacts on the environment entirely (SCG, 2012). Such sustainable design and construction should conserve natural resources, reduce carbon emissions and minimize the negative impact on the local environment whilst safeguarding the economic and social well–being of the people (LDSA, 2008).

4.3.3 Waste management

Construction operations produce large volumes of waste but the effects can be minimized by adopting a construction site waste management plan. This involves specifying and purchasing materials needed for the project, and at the same time recycling and reusing waste to reduce the amount of waste going to the landfills thereby minimizing landfill costs to the organization (LDSA, 2008). Construction organizations can make a considerable contribution to the sustainable development agenda by reducing the amount of construction waste that is sent to landfills through the adoption of excellent waste management systems on construction projects (WRAP, 2011). Hamid and Kamar (2012) believe that construction organizations can achieve environmental, social and economic sustainability through waste minimisation by reducing, reusing and recycling construction materials. The best sustainable waste management practice means using efficient resources that will produce less waste and where waste is produced, dealing with it more efficiently. The adoption of the waste hierarchy framework for sustainable waste management will minimize waste generation on construction sites (Kibert, 2008; Hwang and Tan, 2010). The waste hierarchy framework is illustrated in Figure 4.2. Construction organizations will benefit from the adoption of good waste management practices in terms of reduced material and disposal costs, meeting planning requirements and lower carbon emissions, resulting in increased competitive advantage to the organization (WRAP, 2011). Opoku (2012) states that many construction organizations in the UK have waste reduction targets that help them measure how they are performing annually. These targets are achieved by ensuring that all projects produce site waste management plans and implementing lean construction programmes that increase recycled content.

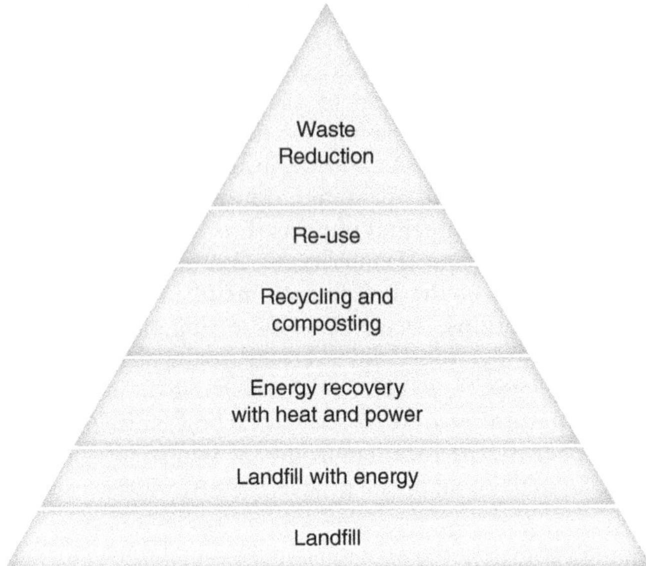

Figure 4.2 Waste Management Hierarchy Framework
(Source: Hwang and Tan, 2010)

4.3.4 Resource and materials

Construction materials have a major impact on the environment and the selection of sustainable materials must strongly be directed to the embodied carbon of the product in order to meet the demands of the future (Carter, *et al.*, 2008). Embodied carbon is described as the amount of carbon released from material extraction, transport, manufacturing, and related activities (Hammond and Jones, 2008). It is argued that sustainable construction costs more than conventional buildings (Yudelson, 2008) because sustainable construction materials cost significantly more than standard (or traditional) materials (Kibert, 2008). However, the use of sustainable resources will help to reduce the operational costs of the building over its whole life. Defects from construction projects and the use of uneconomical construction processes are wasting expensive environmental resources whilst posing danger to both construction workers and the building occupants (Reffat, 2004). Equipment used on construction sites should be energy efficient with low carbon emission achieved by sound maintenance practices. This should include the vehicles that transport materials and personnel to and from construction sites (LDSA, 2008). Selecting the most appropriate materials for a construction project is key and careful consideration should be given to the environmental impact of the materials selected for use, taking into account the source of these materials and health impacts associated with their manufacture and disposal, as well as future maintenance requirements and cost (Hwang and Tan, 2010). When choosing construction materials, designers should consider the lowest resource inputs, environmental emissions and waste throughout the life cycle of the material, as well as possible reuse and recycling options. Opoku (2012) argues that many construction organizations in the UK should source materials which are approved and certified by the appropriate certifying bodies as environmentally friendly: for example, using Forest Stewardship Council (FSC) certified timber and the British standard framework for the responsible sourcing of construction products (BES 6001) certified materials in the construction industry.

4.3.5 Corporate social responsibility and community

It is essential that construction organizations sustain their corporate social responsibility (CSR) to the communities in which they operate whilst also supporting their employees. It is argued that organizations in the UK construction industry have long been implementing the fundamental principles of CSR in the communities in which they operate, providing financial donations, sponsorships, training and other benefits (Barthorpe, 2010). In a qualitative study involving interviews with 85 members of boards of directors, suppliers, employees, customers and community representatives of 17 large corporations in the Australian construction industry, Petrovic–Lazarevic (2008:94) defined CSR as:

> *"A set of principles established by an organization to meet societal expectations of appropriate business behaviour and the achievement of best practice through social benefits and sustainable competitive advantage".*

Further elaboration is provided by Constructing Excellence (2004), which adds that CSR in construction involves dedication to incorporate socially responsible ethics and concerns of stakeholders into day–to–day operations, in a way that fulfils and exceeds current legal and business expectations. An organization's CSR should include its relationships with suppliers, accountability and transparency, sustainability requirement, commitment to local community, moral obligation to be a good citizen and relationships with employees and unions (Porter and Kramer, 2006; Yadong, 2007). Moreover, in the Constructing Excellence report (2004), it was identified that the core values of CSR were transparency, fairness, inclusiveness, responsiveness, integrity, diversity and accountability. As part of construction organizations' CSR, noise impacts associated with both the transportation and use of materials should be reduced drastically. The use of modern construction methods such as the use of framed construction and pre–fabricated components can minimize disruption and noise level (LDSA, 2008). Ghazali (2007) argues that organizations will have an improved corporate image when they practice a more socially responsible business that demonstrates awareness of environmental and ethical issues. A construction organization like any other organization is urged to contribute positively to the development of the society and communities in which they operate. Opoku (2012) pointed out that UK construction organizations recognize that they have a moral obligation to society by offering training schemes for young offenders in the communities in which they operate as a business. Other organizations also provide volunteer man-hours to employees to work for their chosen charities in the communities.

4.3.6 Whole life costing

Whole life costing (WLC) in the construction industry has become particularly pertinent with the increasing focus on the need to reduce carbon emission and enhance sustainable construction (Opoku, 2013). The application of WLC within the construction industry is rapidly increasing due to largely acknowledged benefits that WLC based decision-making can bring to the design and operation of built assets (Richard, 2005; Opoku, 2013). However, the current practice of clients frequently concentrating on initial capital costs and focusing on the lowest tender price rather than the best value is affecting the effective implementation of WLC in the construction industry (Wong, 2010). The Egan report (Egan, 1998) recommended that design should encompass whole life costs, including cost of energy consumption and maintenance. Whole life costing is valuable for comparing alternative building designs, enabling operational costs and benefits to be evaluated against any initial cost increase (Cole and Sterner, 2000; Opoku, 2013). Sev (2009) believes that in order for a construction organization to effectively contribute towards the sustainable development agenda, it should consider the whole life cost of the built asset, impact of construction business activities on the environment and the health and comfort of end–users. Edwards *et al.* (2000) pointed out that construction organizations benefit from demonstrating to their clients an effective, reliable and cost–conscious method of assessing the sustainability of different building options. Moreover, the client then benefits from being offered a clear technique to assess the choices based on financial and environmental criteria could be made. BRE (2004) believe that clients who own and manage buildings on a long term basis want to know their costs of ownership before being committed to any particular alternative design or building.

4.4 Summary

Construction organizations in the UK are actively engaged in sustainability practices such as sustainable procurement, waste management, efficient use of resources and materials, whole life costing, corporate social responsibility and community engagement, sustainable design and carbon reduction commitment. Sustainability is rapidly moving up the agenda as a procurement issue because it is recognized that embedding sustainability into the construction procurement process can bring social, economic and environmental benefits to the organizations. The design of sustainable construction projects may be cheaper to run and can provide safer and healthier living environments. Sustainable design and construction seeks to reduce carbon emissions and negative impacts to the environment. Environmental considerations should be taken into account in the selection of materials and they should be reasonably obtained from sustainable sources. All waste from construction sites should be disposed of properly to licensed tips. The best sustainable waste management practice involves the use of efficient materials that will produce less waste and the management of waste that comes from construction projects more efficiently. Socially, construction organizations can make positive contributions to the communities they operate in and can be organizations which clients and subcontractors want to work with. Even though profit is vital for every organization's economic survival, construction organizations should ensure that they seek to reduce waste, make efficient use of resources and consider the whole life cost effects of all business decisions.

References

Addis, B. and Talbot, R. (2001), "Sustainable construction procurement: A guide to delivering environmentally responsible projects", CIRIA C571, London: Construction Industry Research And Information Association (CIRIA).

Baloi, D. (2003), "Sustainable construction: Challenges and opportunities", In: Greenwood, D. J. (Ed.), *19th Annual ARCOM Conference*, 3–5 September 2003, University of Brighton, Association of Researchers in Construction Management, Vol. 1, pp. 289–97.

Barthorpe, S. (2010), "Implementing corporate social responsibility in the UK construction industry", *Property Management*, Vol. 28 No.1, pp. 4–17.

Beheiry, S. M. A, Chong, W. K. and Haas, C. T. (2006), "Examining the business impact of owner commitment to sustainability", *Journal of Construction Engineering and Management,* Vol. 132 No. 4, pp. 384–92.

Berry, C., and McCarthy, S. (2011), "Guide to sustainable procurement in construction", CIRIA C695, London: Construction Industry Research and Information Association (CIRIA).

BRE (2004), "Whole life costing", Watford: British Research Establishment (BRE).

Brundtland, G. H. (1987), *Our common future: Report of the World Commission On Environment and Development*, Oxford: Oxford University Press.

Carter, K. and Fortune, C. (2003), "Procuring sustainable projects: A grounded approach", In: Greenwood, D. J. (Ed.), *19th Annual ARCOM Conference,* 3–5 September 2003, University of Brighton, Association of Researchers in Construction Management, Vol. 2, pp. 755–64.

Carter, K., Essa, R. and Fortune, C. (2008), "Towards process mapping the development of sustainable housing projects in the UK", In: *Proceedings of International Conference on Building Education and Research (BEAR) "Building Resilience",* Sri Lanka.

Cole, R. J. and Sterner, E. (2000), "Reconciling theory and practice of life cycle costing", *Building Research and Information,* Vol. 28, No. 5-6.

Constructing Excellence (2004), *Corporate Social Responsibility* [Online] available at: http://www.constructingexcellence.net/download/social_responsibility.pdf (Accessed on 22 May 2012).

DEFRA (2006), *Procuring the future: Sustainable procurement national action plan: Recommendations from the Sustainable Procurement Task Force*, London: Department for Environment, Food and Rural Affairs.

DETR (2000), *Building a better quality of life: A strategy for more sustainable construction*, London: Department of the Environment, Transport and Regions.

DTI (2006), "Sustainable construction strategy report 2006", London: Department of Trade and Industry (DTI).

Edum–Fotwe, F. T. and Price, A. D. F. (2009), "A social ontology for appraising sustainability of construction projects and developments", *International Journal of Project Management*, Vol. 27 No. 4, pp. 313–22.

Edwards, S., Bartlet, E. and Dickie, I. (2000), "Using whole life costing and life–cycle assessment for sustainable building design", *BRE Digest 452*, Watford: BRE Press.

Egan, J. (1998), *Re–thinking construction: Report of the Construction Industry Task Force*, London: DETR.

Fairclough, J. (2002), *Rethinking construction innovation and research: A review of government R and D policies and practices*, [Online], Available at: http://www.cidb.org.za/Documents/KC/External_Publications/ext_pubs_fairclough_report.pdf (Accessed 4 December, 2010).

Fenner, R., Ainger, C., Cruickshank, H., and Guthrie, P. (2006), "Widening engineering horizons: Addressing the complexity of sustainable development", *Engineering Sustainability*, Vol. 159 No. 4, pp. 145–54.

Ghazali, N. A. M. (2007), "Ownership structure and corporate social responsibility disclosure: Some Malaysian evidence", *Corporate Governance*, Vol. 7 No. 3, pp. 252–66.

Hamid, Z. A. and Kamar, K. A. M. (2012), "Aspects of off–site manufacturing application towards sustainable construction in Malaysia", *Construction Innovation: Information, Process, Management,* Vol. 12 No. 1, pp. 4 –10.

Hammond, G. P. and Jones, C. I., (2008), "Embodied energy and carbon in construction materials", *Proceedings of the Institution of Civil Engineers – Energy,* Vol. 161 No. 2, pp. 87–98.

Holton, I., Glass, J. and Price, A. (2008), "Developing a successful sector sustainability strategy: Six lessons from the UK construction products industry", *Corporate Social Responsibility and Environmental Management*, Vol. 15 No. 1, pp. 29–42.

Hwang, B. G. and Tan, J. S. (2010), "Green building project management: Obstacles and solutions for sustainable development", *Sustainable Development*, Vol. 20 No. 5, pp. 335–49.

Jones, T., Shan, Y. and Goodrum, P. M. (2010), "An investigation of corporate approaches to sustainability in the US engineering and construction industry", *Construction Management and Economics*, Vol. 28 No. 9, pp. 971–83.

Kibert, C. J. (2007), "The next generation of sustainable construction", *Building Research and Information*, Vol. 35 No .6, pp. 595–601.

Kibert, C. J. (2008), *Sustainable construction: Green building design and delivery.* Hoboken, NJ: Wiley.

Latham, S. M. (1994), *Constructing the Team: Report of the government/industry review of procurement and contractual arrangements in the UK construction industry*. London: HMSO.

LDSA (2008), *Architect's guide to sustainable design and construction*. London: London District Surveyors Association.

Manoliadis, O., Tsolas, I. and Nakou, A. (2006), "Sustainable construction and drivers of change in Greece: A Delphi study", *Construction Management and Economics*, Vol. 24 No. 2, pp. 113–20.

Miller, K. (2010), "Sustainability: Commit in four ways", *Leadership Excellence*, Vol. 28 No.11, p. 8.

Myers, D. (2005), "A review of construction companies attitudes to sustainability", *Construction Management and Economics*, Vol. 23 No.8, pp.781–85.

ODPM (2003), *Sustainable communities: Building for the future*, Office of the Deputy Prime Minister, [online], available at: www.odpm.gov.uk, (Accessed on: 23 October, 2010)

Opoku, A. (2012), "Promoting sustainability practices through leadership within UK construction organizations", (*Unpublished Doctoral Thesis*), University of Salford, Manchester–Salford.

Opoku, A. (2013), "The application of whole life costing in the UK construction industry: Benefits and barriers", *International Journal of Architecture, Engineering and Construction*, Vol. 2 No. 1, pp. 35–42.

Opoku, A. and Ahmed, V. (2013), "Understanding sustainability: A view from organizational leadership within UK construction organizations", *International Journal of Architecture, Engineering and Construction*, Vol. 2 No.2, pp. 133–43.

Parkin, S. (2000), "Context and drivers for operationalizing sustainable development", *Proceedings of ICE*, Vol. 138 No. 6, pp. 9–15.

Petrovic–Lazarevic, S. (2008), "The development of corporate social responsibility in the Australian construction industry", *Construction Management and Economics*, Vol. 26 No. 2, pp. 93–101.

Pitt, M., Tucker, M., Riley, M. and Longden, J. (2009), "Towards sustainable construction: Promotion and best practice", *Construction Innovation*, Vol. 9 No. 2, pp. 201–24.

Porter, M. E. and Kramer, M. R. (2006), "Strategy and society: The link between competitive advantage and corporate responsibility", *Harvard Business Review*, Vol. 84 No.12, pp. 78–92.

Pulaski, M. H. (2004), *Field guide for sustainable construction*. Partnership for Achieving Construction Excellence, The Pennsylvania State University.

Reffat, R. (2004), "Sustainable construction in developing countries", In: *Proceedings of First Architectural International Conference*, Cairo University, Egypt.

Riedy, C. (2003), "A deeper and wider understanding of sustainable development", Institute for Sustainable Futures, University of Technology, Sydney, Australia.

Richard, J. K. (2005), "Re-engineering the whole life cycle costing process", *Construction Management and Economics*, Vol. 23 No. 1, pp. 9–14.

Rohracher, H. (2001), "Managing the technological transition to sustainable construction of buildings: A socio–technical perspective", *Technology Analysis and Strategic Management*, Vol. 13 No.1, pp.137–50.

SCG (2012), "Guidance for project sponsors and project managers guidance note 7: Sustainable design in the built environment", Sustainable Construction Group, Department of Finance and Personnel.

Sev, A. (2009), "How can the construction industry contribute to sustainable development? A conceptual framework", *Sustainable Development*, Vol. 17 No. 3, pp. 161–73.

Shafii, F., Ali, Z. A. and Othman, M. Z. (2006), "Achieving sustainable construction in the developing countries of Southeast Asia", *Proceedings of the 6th Asia–Pacific Structural Engineering and Construction Conference* (APSEC 2006), 5–6 September 2006, Kuala Lumpur, Malaysia.

Singh, T. P. (2007), "Sustainable construction", *Sustainability Tomorrow*, CII-ITC Centre for Excellence in Sustainable Development, Delhi.

Son, H., Kim, C., Chong, W. K, and Chou, J-S. (2011), "Implementing sustainable development in the construction industry: Constructors' perspectives in the US and Korea", *Sustainable Development*, Vol. 19 No. 1, pp. 337–47.

Sourani, A. and Sohail, M. (2011), "Barriers to addressing sustainable construction in public procurement strategies", *Proceedings of the Institution of Civil Engineers, Engineering Sustainability*, Vol. 164 No. 4, pp. 229–37.

UNDESA, (2010), "Buildings and construction as tools for promoting more sustainable patterns of consumption and production", In: Taipale, K. (Ed.), Marrakech Task Force on Sustainable Buildings and Construction, *Sustainable Development Innovation Briefs, Issue 9*, March 2010, New York: United Nations Department of Economic and Social Affairs, Available at: http://www.un.org/esa/sustdev/publications/innovationbriefs. [Accessed: 2 September, 2011].

Williams, D. and Sutrisna, M. (2010), "An evaluation of the role of facilities managers in managing sustainability and remedial actions in reducing CO2 emissions in the built environment", *The Construction, Building and Real Estate Research Conference of the Royal Institution of Chartered Surveyors* (COBRA), Held at Dauphine Universite, Paris, 2–3 September 2010.

Wong, I. L. (2010), "Whole life costing: Towards a sustainable built environment", In: *Responsive Manufacturing-Green Manufacturing (ICRM), 5th International Conference on Responsive Manufacturing*, pp. 248–56,

WRAP (2011), *Achieving good practice waste minimisation and management: Guidance for construction clients, design teams and contractors*. Oxon: Waste and Resources Action Programme.

Yadong, L. (2007), *Global Activities of Corporate Governance*. Oxford: Blackwell Publishing.

Yates, A. (2003), "Sustainable buildings: benefits for constructors", *BRE Information Paper, IP 13/03, Part 4*. Watford: Building Research Establishment.

Yudelson, J. (2008), *Green building revolution*. Washington, DC: Island Press.

5 Drivers and challenges to the adoption of sustainable construction practices

Alex Opoku and Vian Ahmed

5.1 Introduction

The quest for sustainability has put enormous pressure on the construction industry from the government and the general public to improve on its currently unsustainable pattern of project delivery (Adetunji *et al.*, 2003). The construction industry has to change its practices and embrace recycling and reuse of materials as well as the reduction in energy and natural resources use (Baloi, 2003). The construction industry is at the very heart of the challenge we face in the transition to a more sustainable economy. It calls for committed organizations in developing businesses with sustainable products and services. There is now a wide recognition that the construction industry has a vital contribution to make towards sustainable development. However, Leiper *et al.* (2003) comment that the construction industry is also slow in adopting sustainable approaches to its construction project practices. The construction industry has a significant social responsibility to minimize the damage its projects do to the social environment. As a result the UK government has recently introduced a policy of making sustainability project criteria for all publicly founded projects (Essa and Fortune, 2006). However, construction organizations are facing a challenge in attempting to integrate social, economic and environmental responsibility issues in their strategic business plans (Suresh *et al.*, 2008). Construction organizations need to provide the collective vision, strategy and direction towards the common goal of a sustainable future. It is important that such organizations have both the ability as well as the sustainability knowledge to effectively move towards sustainability.

5.2 Sustainability in the construction industry

The concept of sustainable development was coined in 1987 in a report chaired by the Norwegian Prime Minister Gro Harlem Brundtland and commissioned by The United Nations' World Commission on Environment and Development, *Our Common Future* (Brundtland, 1987). The Brundtland Report defines sustainable development by addressing the concept of need and describes it as the process of addressing the needs of the present without compromising the ability of future generations to meet their own needs (Brundtland, 1987). The UK construction industry contributes over 8 percent of GDP and employs 2.1 million people, so this makes it a key sector to spearhead the drive towards sustainable development (DTI, 2006). The UK government is committed to sustainable development and is therefore encouraging the construction industry to change its traditional methods of construction and adopt more sustainable practices (Zhou and Lowe, 2003; Myers, 2005). Over the last twenty years, the concept of sustainability has been growing in importance and it has currently become the basis of most socio–economic activities and developments in the built environments (Edum–Fotwe and Price, 2009). Sustainable construction is conceptualized as having three broad dimensions: social equity, environmental protection and economic growth as a reflection of those of sustainable development. Social sustainability deals with legal, moral and ethical obligations of construction organizations to their stakeholders. Environmental sustainability, on the other hand, addresses the impact of construction activities on the environment by minimizing waste and using natural resources and energy efficiently. Economic sustainability, however, involves improved project delivery resulting in high productivity to maintain a high and stable level of economic growth (Parkin *et al.*, 2003). Construction organizations are currently engaged in the sustainability debate and are formulating business strategies to respond to the increasing demand from governments and the wider public for sustainable construction products (Zhao *et al.*, 2012).

5.3 Sustainable construction drivers

The construction industry has been identified as a key driver in the delivery of UK national carbon emission targets and the wider principles of sustainable development. The industry is therefore being challenged by evolving policy and building standards to change its unsustainable products and practices to align with a more sustainable construction practices (Thomson and El–Haram, 2011). A study by Renukappa *et al.* (2012) to capture the general perceptions of the UK industrial sectors on the concept of sustainability using a qualitative approach confirmed that a growing interest towards sustainability has gathered momentum in the recent time due to global financial crisis. The strong business case for sustainability has boosted the level of interest of governments, stakeholders, investors and consumers. Construction organizations, like all other organizations, can no longer ignore sustainability issues (Holton *et al.*, 2010). Adopting sustainable construction practices will ensure the elimination of negative impacts of construction on the environment and its occupants through efficient use of resources (Jorgensen *et al.*, 2004).

Hunter and Kelley (2007) believe that sustainable construction practices (sustainable design, procurement, waste management, whole life costing, resource use, corporate social responsibility or CSR etc.) could be regarded as one of the main success factors for increasing business competitiveness and efficiency in the industry. To gain competitive advantage, sustainability principles must be integrated into the core processes of an organization such as the organization's project management processes (Gareis *et al.*, 2011). Some of the key drivers for the adoption of sustainable construction include financial incentives, legislation/regulations and client demand (Zhou and Lowe, 2003).

Generally, sustainable construction helps to reduce energy usage and operational costs and fosters protection of natural and social environments, providing healthy and comfortable living environments and also providing economic success for both developers and occupiers. Sustainability enhances building performance, resulting in occupants' and users' improved well–being and productivity (Zhou and Lowe, 2003; Hakkinen and Belloni, 2011). The perception that sustainable buildings cost more is actually not true as the whole life cost of such buildings is cheaper. Sustainability should not be an add-on or a bolt-on decision but rather an integral part of the whole project cycle (Eley, 2011). The Southeast Centre for the Built Environment (SECBE) highlighted in their report "An introductory guide to best practices in construction" that the adoption of sustainable construction practices in construction organizations will increase profitability and competitiveness, provide greater satisfaction, well–being, and added value to customers and users, respect employees and the wider community, enhance site and welfare conditions, protect the natural environment, minimize the consumption of natural resources and energy, reduce waste and avoid pollution during the construction process (SECBE, 2009).

There is now an increasing pressure from shareholders, investors and the public for construction organizations to provide public reports on social and environmental issues relating to their business operations. Organizations are being urged to integrate social and environmental considerations into their working practices and adopt environmental management systems. There is also an appreciation that providing sustainable buildings for people to work and live with high quality public spaces and amenities creates value and will lead to higher investment returns for developers (LDSA, 2008). Sustainability at the organizational level refers to meeting social and environmental needs in addition to the firm's profitability (Porter, 2008). Sustainability makes good business sense because it is of increasing importance to the efficient, effective and responsible operation of business. Some of the primary drivers towards the adoption of more sustainable business practices in construction organizations in UK are government policy or legislation, reputation as a green company and gaining competitive advantage (Bennett and Crudgington, 2003; Woodall *et al.,* 2004; Holton *et al.*, 2008). Even though Griffith and Bhutto (2008) acknowledged that cost of building remains the key factor for construction organizations' competitiveness, the authors argue that it is essential for construction organizations to consider environmental aspects of their business in order to remain competitive. Economically, sustainable construction provides capital cost savings, reduced running and maintenance costs, tax savings, added value, more efficient resource use, productivity improvement, increased organizational effectiveness, increased investment returns, the generation of positive image and support for the local economy (Yates, 2001; Zhou and Lowe, 2003; Al-Yami and Price, 2006). Additionally, benefits experienced by organizations adopting sustainability include reduced costs, fewer construction defects, increased market share, improved health and safety and an improved corporate image. Organizations implementing sustainable practices can become the companies of first choice. An improved image is economically important in attracting work from repeat and new clients (Yates, 2003). Sustainable construction buildings offer lower cost in terms of energy used, waste disposal, water, environmental and emissions costs, operations and maintenance costs and savings from increased productivity and health (Kats *et al.*, 2003). The above discussion therefore illustrates the economic benefits of sustainability to construction business clients, customers and end–users.

A seminal research project (Partners in Innovation–PII) by the Building Research Establishment (BRE) and supported by the Department of Trade and Industry (DTI)'s construction directorate and industry stakeholders studied the business benefits of sustainable construction practices. The results indicated that construction organizations engaged in sustainable practices benefit from increased productivity, reduced material costs, reduced wastage, reduced landfill tax burdens and transport costs (Yates, 2003). Environmentally, sustainable construction practices preserve the natural environment (Zhou and Lowe, 2003) and contribute to environmental protection (Baloi, 2003) improved air and water quality, and minimal waste disposal, with less use of energy and water (Al-Yami and Price, 2006). Socially, adopting sustainable practices will support the local economy through the provision of employment (Zhou and Lowe, 2003), staff recruitment and retention (Yates, 2001) and improvement of staff working conditions (Baloi, 2003). Promoting the adoption of sustainable construction practices will bring long–term benefits to the built environment and its occupants in particular (Al-Yami and Price, 2006) and the national economy as a whole through a reduction in carbon emissions and the use of natural resources (Hakkinen and Belloni, 2011). Sustainable construction practices will enhance efficiency, protect the environment, gain competitive advantages and achieve value for money (Bennett and Crudgington, 2003). Arif *et al.* (2009) add that some of the key drivers behind the adoption of sustainable construction include regulation, green corporate image, corporate social responsibility (CSR) and cost savings through reduced cost of energy and waste. The benefits of adopting sustainable construction practices are many but some of the key benefits are the efficient use of resources, the provision of an enhanced working, living, and learning environment and reduction in the building's impacts on human health and the environment. The United States Green Building Council (USGBC) believes that occupants of sustainable buildings save cost in terms of energy, water, maintenance and operations, thereby enhancing their productivity (Kats *et al.*, 2003). The influential sustainable construction drivers and literature sources are shown in Table 5.1.

Table 5.1 Sustainable construction drivers identified in literature

Sustainable Construction Drivers	*Literature Source*
Resource Efficiency	Yates, 2003; Jorgensen *et al.*, 2004; Al-Yami and Price, 2006; Arif *et al.*, 2009
Competitive Advantage	Hunter and Kelley, 2007; Gareis *et al.*, 2011
Legislation/legal requirement	Zhou and Lowe, 2003; Bennett and Crudgington, 2003; Woodall *et al.*, 2004; Holton *et al.*, 2008; Arif *et al.*, 2009
Reputation/Image	Bennett and Crudgington, 2003; Yates, 2003; Zhou and Lowe, 2003; Woodall *et al.*, 2004; Al-Yami and Price, 2006; Holton *et al.*, 2008; Arif *et al.*, 2009
Client demand	Zhou and Lowe, 2003
Win more clients/Financial incentives	Yates, 2003; Zhou and Lowe, 2003; Al-Yami and Price, 2006
Attract and retain good employees	Yates, 2001

5.4 Sustainable construction challenges

Despite the numerous benefits claimed to be associated with sustainable construction, adopting such practices has its own challenges. Thomson and El-Haram (2011) explained that, delivering sustainability in practice remains a challenge partly due to a range of traditional cultural and structural barriers such as the lack of integration between the different project stages and professions in the project team. Zhou and Lowe (2003) argue that the promotion of sustainable construction faces several economic challenges due to poor understanding of economic benefits that can be achieved. There is lack of understanding of the business case for sustainability and lack of building regulations and planning policy that enforce sustainable construction (Pitt *et al.*, 2009). Even though there are high investment costs for sustainable construction projects compared with traditional building practices (Hakkinen and Belloni, 2011), however, this can be addressed by utilizing the whole life cycle costing technique, moving from cost to value and from short–term to long–term cost perspectives (Al-Yami and Price, 2006). Additional construction cost has been cited by many researchers as the most widespread barrier to the implementation of sustainable construction (Sodagar and Fieldson, 2008; Hakkinen and Belloni, 2011).

The misconception that sustainable construction will cost more and the lack of a visible market value discourages both construction organizations and investors from fully embracing its principles in their organizational practices. This common perception that sustainable construction is more expensive in terms of capital costs compared to normal mainstream buildings is a big challenge to the adoption of sustainable construction (Zhou and Lowe, 2003).

In a qualitative study by Williams and Dair (2007) involving five case studies of completed developments in England, it was identified that there were a number of barriers to sustainable construction practices, including lack of consideration of sustainability measures by stakeholders, sustainability not being required by clients, real and perceived costs and inadequate expertise and powers. The complex and fragmented nature of the construction industry is suspected to be another reason why there is a tendency to resist changes leading towards sustainability. Sustainable construction projects require close working interactions with all the project team from design to completion stages (Riley *et al.*, 2003; Hakkinen and Belloni, 2011). Most construction organizations concerned with the implementation of sustainable practices have the perception that it will result in increased risks, higher capital costs, even difficulties in obtaining financial support and the lack of awareness of market value (Zhou and Lowe, 2003). The construction industry is client–driven and therefore plays a major role in the adoption of sustainable construction. The challenge is that lack of client awareness and demand for sustainable building (Pitt *et al.*, 2009) will affect the agenda towards sustainability. Currently, major clients such as governmental and local authority organizations are significantly impacting the adoption of sustainable construction practices (Hakkinen and Belloni, 2011). Construction organizations should therefore be proactive in offering sustainable services and products to prospective clients (Berry and McCarthy, 2011).

A critical evaluation of the human resource capacity of the construction industry shows that the industry lacks the human resource capacity in terms of both numbers and the skills level of professionals, tradesmen and labourers to support the implementation of sustainable construction. A study by Alkhaddar *et al.* (2012) involving a survey with 133 office and site–based construction workers on their awareness and understanding of sustainability in their working environment revealed that most construction industry workers seem keen on sustainability issues; however, few people feel they have the adequate skills to do their job and only do the minimum required with regards to sustainability practices. There are not enough trained human resources with the required skills to perform sustainable construction operations (Reffat, 2004). In a case study by Mukherjee and Muga (2010) that developed an integrative framework for studying sustainable practices and their adoption in the architecture, engineering, and construction (AEC) industries, the researchers identified a lack of a collective theory to promote a systems approach to sustainable methods as one of the major challenges facing the adoption of sustainability practices in the AEC industry. However, Hamid and Kamar (2012) add that as sustainability is continuing to be treated as a discrete problem with an isolated solution, it will be difficult to blend it into construction processes. It is therefore important that the adoption of sustainability practices in construction organizations is led by committed leadership within these organizations. Table 2 shows the significant challenges facing the adoption of sustainable construction practices in the construction industry.

Table 5.2 Sustainable construction challenges identified in literature

Sustainable Construction Challenges	*Literature Source*
Lack of understanding sustainability	Zhou and Lowe, 2003
Increase cost	Al-Yami and Price, 2006; Sodagar and Fieldson, 2008; Hakkinen and Belloni, 2011
Perception that sustainability cost more	Zhou and Lowe, 2003; Williams and Dair, 2007
Contract requirement/ procurement practices	Adetunji *et al.*, 2008; Sodagar and Fieldson, 2008; Hakkinen and Belloni, 2011
Client demand and understanding of sustainability	Pitt *et al.*, 2009
Lack of skilled tradesman for sustainable construction	Reffat, 2004

5.5 Sustainable construction in the UK construction industry

The UK government has put in place several initiatives to encourage reform in the construction industry through UK construction industry through the design for minimum waste and energy use, conservation of resources and respect for people and the local environment (DETR, 2000). This section of the chapter presents the result of a study by Opoku (2012) on the drivers and challenges facing the effective implementation of sustainable construction practices in UK construction organizations. The study adopted a mixed methods research approach that provided both qualitative and quantitative evidence from 15 interviews and 200 surveys, which gives a more complete picture of the drivers and challenges facing the promotion and implementation of sustainability practices in construction project delivery.

5.5.1 Drivers

The promotion of sustainable project delivery practices in most construction companies is driven by either government policy or regulations or revenue and income. The construction industry is client–driven and therefore a client requirement for sustainability or lack of it could be a key driver or a major challenge on a project. Sustainability is good for business, the client and the environment; having a green reputation will help construction organizations win more business and attract and retain good graduates into the company (Opoku, 2012). In a qualitative study by Opoku and Fortune (2013) involving interviews with 15 sustainability leaders in UK construction organizations charged with the promotion of sustainable construction practices, a number of sustainable construction drivers were identified including legislation and legal requirements; cost efficiencies; waste issues; social issues; ethics; stakeholder influence; retention of key staff; competitiveness; green agenda; resource depletion etc. These drivers are underpinned by a moral obligation to act with integrity and do the right thing.

Factors such as reputation, winning more business, competitive advantage, reducing cost, attracting and retaining best staff are the common factors driving the promotion of sustainability practices in both contactor and consultant organizations. There are other factors which are peculiar to either contactor or consultant organizations. For example, customer satisfaction, revenue and income and government policy are factors driving consultant organizations in the construction industry (Opoku, 2012; Opoku and Fortune, 2013). Similarly, drivers such as moral/ethical issues, client requirements and minimizing risk only drive contractor organizations. It is therefore important to note that while there are common drivers to contractor and consultant construction organizations, differences in organizational business activities and strategies have resulted in other factors which are only pertinent to either type of organization. Figure 5.1 shows key sustainable construction drivers identified in the UK construction industry through the study described above.

5.5.2 Challenges

Construction organizations face challenges in an attempt to effectively implement sustainable construction practices. One major challenge to the adoption of sustainability is the real or perceived cost associated with sustainability. The perception that sustainability costs more is a real challenge.

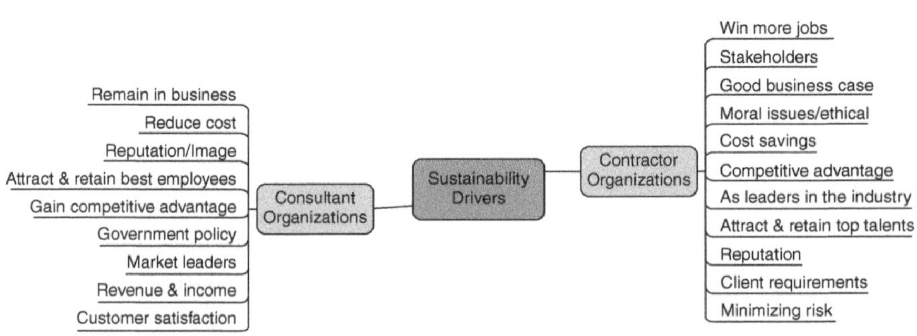

Figure 5.1 Sustainable construction drivers

Most clients only look at the initial capital cost without considering the whole life cost of the built asset; the perception that sustainability costs more still exists. Clients need to understand the benefits of whole life costing in order to make an informed decision when adopting sustainability (Opoku, 2012; Opoku and Ahmed, 2014). Knowledge and understanding of sustainability in the construction industry is crucial if the promotion of sustainability practices in construction organizations is to be successful. Sustainability involves environmental, social and economic impacts of our business activities. There is lack of knowledge and skills of employees who should be involved in the implementation of sustainability practices. Construction organizations lack trades people and operatives with the right skills and knowledge in sustainability to engage in the promotion and implementation of sustainability practices in the delivery of construction projects (Opoku and Fortune, 2013; Opoku and Ahmed, 2014).

The promotion of sustainability practices in large organizations is a huge challenge. It is sometimes difficult for sustainability policies to trickle down in organizations with large company structure and diversity of business services. Each department within the organization may have different priorities and conflicting targets and managing such diversity with sustainability at the core of the organization's strategy is very challenging for intra-organizational leaders. Getting other departments to understand the business case for sustainability as well as keeping every business unit on track is very difficult (Opoku, 2012; Opoku and Fortune, 2013; Opoku and Ahmed, 2014).

In order to successfully promote and implement sustainability practices in construction organizations, it is important that the board of the company share the sustainability vision. The challenge is that, sometimes a company board lacks the awareness and the knowledge to make decisions on sustainability. Senior members of company boards have other high priorities and sustainability is at the bottom of most company boards' priority list (Opoku, 2012; Opoku and Ahmed, 2014). Notwithstanding the above, middle management professionals such as project managers and site managers lack the knowledge, skills and understanding required to support the promotion and implementation of sustainability practices in the delivery of construction projects. Such professionals sometimes feel sustainability is an added luxury and only see a successful construction project as the one that is completed on time, on budget and of good quality. However, sustainability is currently a key benchmark in evaluating a successful construction project (Opoku, 2012; Opoku and Ahmed, 2014).

Finally, the procurement process in the construction industry is another challenge. The type and forms of contracts used in the construction industry do not encourage the promotion of sustainability practices. The culture of awarding construction contracts to the lowest bidder does not facilitate the proactive adoption of sustainability. Externally, clients are bound by contractual requirements that are based on lowest price and not sustainability principles (Opoku, 2012). Both contractor and consultant

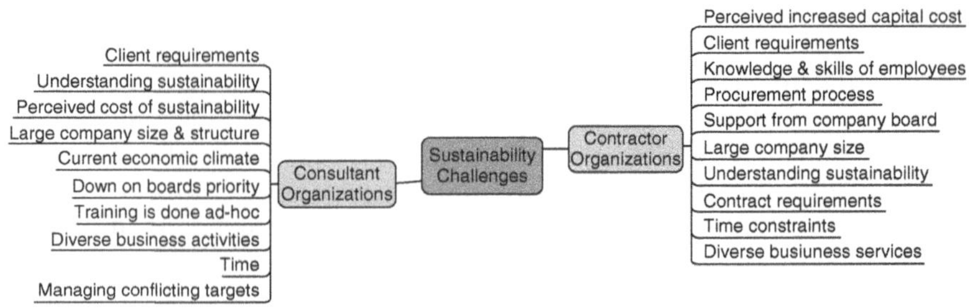

Figure 5.2 Sustainable construction challenges

organizations in the UK construction industry share common challenges such as client requirements, company size and diversity in business activities, cost associated with sustainability, what sustainability really means, support from company board and time constraints. However, there are other challenges facing either contractor organizations or consultant organizations. Factors such as contract requirements, procurement process and knowledge and skills of employees are challenges facing the adoption of sustainability practices in contractor organizations in the UK. However, the management of conflicting business targets and the current economic climate is affecting UK consultant organizations more than contactor organizations when it comes to the promotion of sustainability practices in construction project delivery. Figure 5.2 shows key sustainable construction challenges facing the UK construction industry, identified through the study described above.

5.6 Summary

Construction organizations believe that there is a good business case to pursue sustainability practices in the industry. Adopting sustainable construction practices will drive out waste in terms of materials, people, resources and energy, reducing cost and eventually making such companies more efficient. The desire to be a leader in sustainability in the construction industry is driving most organizations to promote and adopt sustainability practices in the delivery of construction projects. Construction organizations with the aspirations to pursue sustainability are driven by a number of factors; however, company reputation, competitive advantage and legislation or legal requirements are identified as key factors. In addition, the desire to remain in business is a major driver for most construction organizations in the UK. Construction organizations face challenges in an attempt to effectively implement sustainable construction practices. Common challenges include real or perceived cost, client requirements, contract requirements and lack of understanding sustainability.

References

Adetunji, I., Price, A., Fleming, P. and Kemp, P. (2003), "Sustainability and the UK construction industry: A review", *Proceedings of the Institution of Civil Engineers, Engineering Sustainability,* Vol. 156 No. 4, pp. 185–99.

Alkhaddar, R., Wooder, T., Sertyesilisik, B. and Tunstall, A. (2012), "Deep learning approach's effectiveness on sustainability improvement in the UK construction industry", *Management of Environmental Quality: An International Journal,* Vol. 23 No. 2, pp. 126–39.

Al-Yami, A. M. and Price, A. D. F. (2006), "A framework for implementing sustainable construction in building briefing project", In: Boyd, D. (Ed), *Proceedings of the 22nd Annual ARCOM Conference*, 4–6 September, 2006, Birmingham, UK, Association of Researchers in Construction Management, pp. 327–37.

Arif, M., Egbu, C., Haleem, A., Kulonda, D. and Khalfan, M. (2009), "State of green construction in India: Drivers and challenges", *Journal of Engineering, Design and Technology*, Vol. 7 No. 2, pp. 223–34.

Baloi, D. (2003), "Sustainable construction: Challenges and opportunities", In: Greenwood, D. J. (Ed), *19th Annual ARCOM Conference*, 3–5 September 2003, University of Brighton, Association of Researchers in Construction Management, Vol. 1, 289–97.

Bennett, J. and Crudgington, A. (2003), "Sustainable development: Recent thinking and practice in the UK", *Proceedings of the Institution of Civil Engineers, Engineering Sustainability*, Vol. 156 No. 1, pp. 27–32.

Berry, C., and McCarthy, S. (2011), "Guide to sustainable procurement in construction", CIRIA C695, London: Construction Industry Research and Information Association (CIRIA).

Brundtland, G. H. (1987), *Our common future: Report of the World Commission on Environment and Development,* Oxford: Oxford University Press.

DETR (2000), *Building a better quality of life: A strategy for more sustainable construction*, London: Department of the Environment, Transport and Regions.

DTI (2006), *Review of Sustainable Construction*, London: Department of Trade and Industry.

Edum–Fotwe, F. T. and Price, A. D. F. (2009), "A social ontology for appraising sustainability of construction projects and developments", *International Journal of Project Management*, Vol. 27 No. 4, pp. 313–22.

Eley, J. (2011), *Sustainability building: The client's role*, London: RIBA Publishing.

Essa, R. and Fortune, C. (2006), "Project price forecasting processes and the pre–construction evaluation of sustainable housing associations projects", *22nd Annual ARCOM Conference*, 4–6 September 2006, UCE, Birmingham, Association of Researchers in Construction Management, Vol. 2, 543–52.

Gareis, R., Huemann, M. and Martinuzzi, A. (2011), "What can project management learn from considering sustainability principles?", In: Kankonen, K. and Latvanne, A. (Eds), *Project Perspectives*, Vol. XXXIII, pp. 60–65.

Griffith, A. and Bhutto, K. (2008), "Improving environmental performance through integrated management systems (IMS) in the UK", *Management of Environmental Quality: An International Journal*, Vol. 19 No. 5, pp. 565–78.

Hakkinen, T. and Belloni, K. (2011), "Barriers and drivers for sustainable building", *Building Research and Information*, Vol. 39 No. 3, pp. 239–55.

Hamid, Z. A. and Kamar, K. A. M. (2012), "Aspects of off–site manufacturing application towards sustainable construction in Malaysia", *Construction Innovation: Information, Process, Management,* Vol. 12 No. 1, pp. 4–10.

Holton, I., Glass, J. and Price, A. (2008), "Developing a successful sector sustainability strategy: Six lessons from the UK construction products industry", *Corporate Social Responsibility and Environmental Management*, Vol. 15 No. 1, pp. 29–42.

Holton, I, Glass, J. and Price, A. D. F. (2010), "Managing for sustainability: Findings from four company case studies in the UK precast concrete industry", *Journal of Cleaner Production*, Vol. 18 No. 2, pp. 152–60.

Hunter, K. and Kelly, J. (2007), "Case studies on Scottish construction: Companies embracing sustainability", *The Construction, Building and Real Estate Research Conference of the Royal Institution of Chartered Surveyors (COBRA 2007)*, Georgia Tech, Atlanta, USA, 6–7 September 2007.

Jorgensen, B., Emmit, S. and Bonke, S. (2004), "Integrating design and construction: Upstream and downstream values", *Proceedings of the CIB World Building Congress*, Toronto, Ontario, Canada.

Kats, G., Alevantis, L., Berman, A., Mills, E. and Perlman, J. (2003), "The costs and financial benefits of green buildings", *A Report to California's Sustainable Building Task Force*, Massachusetts Technology Collaborative.

Leiper, Q., Fagan, N., Engstrom, S. and Fenn, G. (2003), "A strategy for sustainability", *Proceedings of the Institution of Civil Engineers, Engineering Sustainability*, Vol. 156 No. 1, pp. 59–66.

LDSA (2008), *Architect's guide to sustainable design and construction*, London: London District Surveyors Association.

Mukherjee, A. and Muga, H. (2010), "An integrative framework for studying sustainable practices and its adoption in the AEC industry: A case study", *Journal of Engineering and Technology Management*, Vol. 27 No. 3, pp. 197–214.

Myers, D. (2005), "A review of construction companies' attitudes to sustainability", *Construction Management and Economics*, Vol. 23 No. 8, pp. 781–85.

Opoku, A. (2012), "Promoting sustainability practices through leadership within UK construction organizations", (*Unpublished Doctoral Thesis*), University of Salford, Manchester–Salford.

Opoku, A. and Ahmed, V. (2014), "Embracing sustainability practices in UK construction organizations: Challenges facing intra–organizational leadership", *Journal of Built Environment Project and Asset Management*, Vol. 4 No. 1, pp. 90–107.

Opoku, A. and Ahmed, V. (2013), "Understanding sustainability: A view from organizational leadership within UK construction organizations", *International Journal of Architecture, Engineering and Construction*, Vol. 2 No. 2, pp. 133–43.

Opoku, A and Fortune, C (2011), "The implementation of sustainable practices through leadership in construction organizations", In: Egbu, C. and Lou, E. C. W. (Eds), *Procs 27th Annual ARCOM Conference,* 5–7 September 2011, Bristol, UK, Association of Researchers in Construction Management, 1145–54.

Parkin, S., Sommer, F., and Uren, S (2003), "Sustainable development: Understanding the concept and practical challenge", *Proceedings of the Institution of Civil Engineers, Engineering Sustainability,* Vol. 156 No. 1, pp. 19–26.

Pitt, M., Tucker, M., Riley, M. and Longden, J. (2009), "Towards sustainable construction: Promotion and best practice", *Construction Innovation*, Vol. 9 No. 2, pp. 201–24.

Porter, T. B. (2008), "Managerial applications of corporate social responsibility and systems thinking for achieving sustainability outcomes", *Systems Research and Behavioural Science*, Vol. 25 No. 3, pp. 397–411.

Reffat, R. (2004), "Sustainable construction in developing countries", *Proceedings of First Architectural International Conference*, Cairo University, Egypt.

Renukappa, S., Egbu, C., Akintoye, A. and Goulding, J. (2012), "A critical reflection on sustainability within the UK industrial sectors", *Construction Innovation: Information, Process, Management*, Vol. 12 No. 3, pp. 317–34.

Riley, D., Pexton, K. and Drilling, J. (2003), "Procurement of sustainable construction services in the United States: The contractor's role in green buildings", UNEP Industry and Environment, *Sustainable Building and Construction*, pp. 66–71.

SECBE (2005), *An introductory guide to best practice in construction*, Reading: South East Centre for the Built Environment.

Sodagar, B. and Fieldson, R. (2008), "Towards a sustainable construction practice", *Construction Information Quarterly*, Vol.10 No. 3, pp. 101–08.

Suresh, S., Egbu, C. O. and Olomolaiye, P. (2008), "Leadership and its role for the successful deployment of knowledge capture initiatives", In: Carter, K., Ogulana, S. and Kaka, A. (Eds), *Transformation through construction–Joint 2008 CIB W065/W055 Symposium Proceedings*, 15–17 November 2008, Dubai, pp. 124–25.

Thomson, C. S. and El–Haram, M. (2011), "Exploring the potential of sustainability action plans within construction projects", In: Egbu, C. and Lou, E.C.W. (Eds), *Procs 27th Annual ARCOM Conference,* 5–7 September 2011, Bristol, UK, Association of Researchers in Construction Management, pp. 1085–94.

Williams, K. and Dair, C. (2007), "What is stopping sustainable building in England? Barriers experienced by stakeholders in delivering sustainable developments", *Sustainable Development*, Vol. 15 No. 3, pp. 135–47.

Woodall, R., Cooper, I., Crowhurst, D., Hadi, M. and Platt, D. (2004), "MaSC: Managing sustainable companies", *Proceedings of the Institution of Civil Engineers, Engineering Sustainability*, Vol. 157 No.1, pp.15–21.

Yates, A. (2001), "Quantifying the business benefits of sustainable buildings: Summary of existing research findings", Centre for Sustainable Construction, Watford: Building Research Establishment (BRE).

Yates, A. (2003), Sustainable buildings: Benefits for constructors, *BRE Information Paper, IP 13/03, Part 4*, Watford: Building Research Establishment.

Zhao, Z., Zhao, X., Davidson, K. and Zuo, J. (2012), "A corporate social responsibility indicator system for construction enterprises", *Journal of Cleaner Production*, Vol. 29 No. 30, pp. 277–89.

Zhou, L. and Lowe D. J. (2003), "Economic challenges of sustainable construction", In: Proverbs, D. (Ed), *Proceedings of the RICS Foundation Construction and Building Research Conference (COBRA 2003)*, 1–2 September 2003, Wolverhampton, UK, The RICS Foundation.

6 Sustainability leaders for sustainable cities

Begum Sertyesilisik and Egemen Sertyesilisik

6.1 Introduction

The growing industrialization and scale of economic activity have transformed the world's resources into wealth, resulting in increases in the consumption level of materials and energy as well as adverse effects on ecosystems and societies (Linnenluecke and Griffiths, 2013). The world's habitat is being deteriorated (i.e. global air pollution problems, loss in biodiversity) (Dittmar, 2014). The data about the state of the world presented at the Rio+20 Summit revealed that today's human beings live less sustainably than they did in 1992 and that economic activities' environmental impacts have the potential for causing uncontrollable global disasters (Dittmar, 2014). The main attempts to sustain the world's habitat include the Kyoto Protocol, which has fuzziness in its goals, and the 2012 UN RIO +20 United Nations Conference on Sustainable Development conference, which revealed failure to move toward sustainability goals (Dittmar, 2014). Most commentators, however, suggest that the Rio +20 conference failed to reconcile the ambitions of sustainable development and poverty eradication (Tukker, 2013). Rio +20 enabled the launch of the Global Research Forum on Sustainable Production and Consumption (GRF–SPaC) ("Rio + 20") (Vergargt *et al.*, 2014).

According to Dittmar (2014), "Our current policies and practices are leading . . . to total collapse and to the extinction of . . . species, our own quite possibly included." Our political global leaders and their institutions continue pursuing the same "sustainable development" policies (Dittmar, 2014), which cannot prevent the deterioration of the world as global warming and increase in the amount of CO2 emissions still continue. It is said that insanity is "doing the same thing over and over again and expecting different results." As human beings dominate the health and well–being of the world and its inhabitants (Cortese, 2003), they need to act more proactively to sustain and regenerate the world's habitat. There is need that:

- technology and economic activities are designed to sustain the natural environment and to enhance human health and well–being;
- technology is inspired by biological models operating on renewable energy;

- waste is eliminated through using waste as a raw material/nutrient for other species/activities or returning it into the cycles of nature;
- human activities restore the biological diversity and complexity of ecosystems;
- humans live off nature's interest;
- professionals understand their connections to nature and to other humans.

There is a need for sustainable cities in order to sustain and regenerate the world's habitat. The world's cities occupy just 2 percent of the earth's land, but account for 60–80 percent of energy consumption and 75 percent of carbon emissions (UN 2014). Shabu (2010) describes the effects of the cities as follows:

> *"The cities exercise enormous control over national economies, they provide jobs, access to the best cultural, educational and health facilities and they act as hub for communication and transports which are necessary conditions for economic development of any nation [but] they also cluster massive demand for energy, generate large quantities of waste and concentrate pollution as well as social hardship".*

There is rapid urbanization. Half of the world's population lives in cities today whereas by 2030, almost 60 percent of the world's people are expected to live in urban areas (UN, 2014). This rapid urbanization puts pressure on fresh water supplies, sewage, the living environment, and public health (UN, 2014) and reveals the importance of sustainable cities which can be achieved by sustainability leaders. A carefully designed and managed human impact can help regeneration of previously damaged areas (Dittmar, 2014). For this reason, this chapter will focus on the concept of sustainability leadership, its impacts on cities' sustainability and the ways to transform individuals in the community into sustainability leaders (i.e. education, mass media).

6.2 Sustainability leaders

Sustainability leaders who act as change agents are the keys for successful transformation towards sustainable cities. Hesselbarth and Schaltegger (2014: 26) define change agents as follows:

> *"Change agents are opinion leaders and driving forces in change processes. . . . A change agent for sustainability is an actor who deliberately tackles social and ecological problems with entrepreneurial means to put sustainability management into organizational practice and to contribute to a sustainable development of the economy and society. . . . Change agents for sustainability are not necessarily senior managers but can be individuals on all levels internal or external of an organization. The*

> *definition of a change agent for sustainability thus also includes those who successfully initiate and promote change toward sustainable development on a lower hierarchy level and without a specific mandate".*

Leadership is crucially important for the establishment and implementation of effective policies needed for the transformation of cities into sustainable ones. Change–oriented leaders (whether or not they possess formal leadership positions) are needed for cities to be sustainable. The society, as a whole, needs to act as agents for change. For the transformation of the cities into sustainable ones, there is need for the transformation of individuals in communities into sustainability leaders, increasing societies' awareness and consciousness for sustainability so that they can embrace the change for sustainability and relevant policies.

Hesselbarth and Schaltegger (2014)'s research revealed that acting successfully as a change agent for sustainability requires flexibility to change/create jobs and a large set of different competencies, including general management and subject-specific competencies in sustainability management, methodological, social, and personal competencies. Transforming ordinary people into sustainability leaders (change agents for sustainability) requires:

- **sustainability communication and social learning:** As social learning enables people to learn from and with each other while communicating and increases their competency, it can encourage communities to adopt sustainability principles, ethics and values (Schmitt, 2011). Communication processes that facilitate learning can be achieved through collaborative learning, creating motives for discussion, and designing learning environments
- **education:** Education needs to be competence–oriented (Hesselbarth and Schaltegger, 2014). Schools at different levels can contribute to raising the environmental consciousness of future generations who can become sustainability leaders in the future and who can take the responsibility for the environment. Posch (1999) describes schools' role in transition towards sustainability at three levels, as follows:
    - o "At the pedagogical level schools aim at creating . . . learning experiences and at involving pupils in ecological ways of thinking, acting and feeling in school, in their family and in the community
    - o "At the social/organisational level schools aims at building and cultivating a culture of communication and decision making
    - o At the technical/economic level schools aim at the ecologically sound and economic use of resources."

Higher education institutions, companies and consumers, politicians and citizens, as well as media can act as sustainability leaders influencing sustainability of cities.

6.2.1 Higher education institutions as sustainability leaders

Education plays an important role in sustainability of the world's habitat. "The crisis of global ecology is . . . a crisis of values, ideas, perspectives, and knowledge, which makes it a crisis of education" (Cortese, 2003). Higher education institutions train and educate future professionals including city planners, architects, and politicians who will play an active role in transformation of the cities into sustainable ones. Furthermore, they train other professionals and help to increase the sustainability consciousness of the future professionals who can spread sustainability principles to all sectors. Higher education institutions have responsibility to increase the awareness, knowledge, skills, and values needed to create a sustainable future (Cortese, 2003). As higher education institutions are leaders in creating a sustainable society, they need to support local and regional communities and lead community improvement, especially through enabling improved learning for all; students' preparation for citizenship and career; cooperation and satisfaction across the university; and fulfilment of higher education's moral as well as social responsibilities.

Jain *et al.*'s study (2013) revealed the impact of sustainability education on students. According to Jain *et al.*, the TERI University's postgraduate program in Environmental Studies and Resource Management motivated and trained students who have taken the initiative for achieving sustainability on the campus through an environmental management plan and policy. Higher educational institutions' role in promoting sustainability has been emphasized in 1990 with the Talloires Declaration (Davis *et al.*, 2003) as well as in the Halifax Declaration (1991). Principle 4 of the Halifax Declaration (1991) requires universities to "enhance their capacity . . . to teach and practice sustainable development principles" (Hancock and Nuttman, 2014). The "American Association of University Leaders for a Sustainable Future, as stated by Clugston, works to identify characteristics of higher educational institutions fully committed to sustainability", state Davis *et al.* (2003). Despite these milestones undertaken, many universities are lagging behind companies in helping societies to become more sustainable (Lozano *et al.*, 2013). Main barriers which prevent higher education institutes to be active in sustainability transition include lack of sufficient staff and adequate resources in educational institutions as well as many academics' perceptions of the "sustainability" concept as too abstract and broad (Leal Filho 2000, as quoted in Davis *et al.*, 2003).

According to Lozano *et al.* (2013), universities can do the following to become sustainability leaders and educate change leaders:

- collaborating with other universities;
- fostering transdisciplinarity;
- integrating sustainable development into their institutional framework;
- creating on-campus life experiences;
- "educating–the–educators";

- understanding and building upon the needs of present and future generations,
- empowering university leaders and staff to implement new paradigms and become more proactive in making sustainable development an integral part of their system;
- enabling collaboration among academics and citizens so that sustainable development's integration into societal decision-making processes is supported;
- accepting the transformation process as a continuous process requiring dedication.

Furthermore, Davis *et al.* (2003) recommend the following actions to be carried out so that universities can become sustainability leaders:

- providing an action plan for implementing sustainability initiatives;
- assessing the extent to which sustainability concepts have been incorporated into the institution's academic functions;
- allocating time and providing support to train faculty and staff to collect classroom materials and to assess progress;
- establishing written mission statements, academic programs, energy and purchasing practices (Clugston 1999 and Leal Filho 1999 as quoted by Davis *et al.*, 2003).

Higher education's effectiveness in creation of sustainable communities can be increased by deep learning and high impact educational practices (i.e. community-based sustainable projects) as supported by O'Brien and Sarkis' (2014)'s research on the usage of community–based sustainability projects as an integral part of a deep learning curriculum offered by management, engineering, and public policy (social science) disciplines in university settings. Furthermore, Al-Khaddar *et al.* (2012)'s study supported the effectiveness of a deep learning approach on sustainability improvement in the construction industry. Higher education institutes tend to integrate sustainability topics into the various disciplines including business management. As cities are economic hubs, the production and management processes need to be carried out based on the sustainability principles so that the cities' habitat is sustained. This can be achieved by sustainability leaders who have studied business management curriculum covering sustainability topics. Hesselbarth and Schaltegger (2014) stated that education for sustainability management needs to have knowledge, skills, and attitudes components (Lozano *et al.*, 2013; Stubbs and Schapper, 2011; WEC and Net Impact, 2011) which can be described as follows (Hesselbarth and Schaltegger, 2014):

- Knowledge: The students need to gain knowledge about ecological concepts, environmental management systems and practices, specific approaches to nature and sustainability, and concepts of social global justice (e.g. Waddock, 2007, as quoted in Hesselbarth and Schaltegger, 2014)

- Skills: The development of advanced communication, negotiation, critical analysis, and change management skills need to be included in postgraduate studies (Hind *et al.*, 2009 as quoted in Hesselbarth and Schaltegger, 2014)
- Attitudes: students need to be encouraged to question their perception of the world and to develop reflective thinking (Rands, 2009; Shephard, 2008 as quoted in Hesselbarth and Schaltegger, 2014).

There is a trend that the business and management curriculum covers sustainability topics such as (O'Brien and Sarkis, 2014) environmental topics important for CSR and corporate sustainability, cleaner production, energy efficiency, eco–design, and design for the environment (Hesselbarth and Schaltegger, 2014). Instead of emphasizing individual learning and competition, resulting in professionals who are ill prepared for cooperative efforts (Cortese, 2003), there is need for transforming education at all levels, emphasizing the importance of sustainability (McIntosh *et al.*, 2001). Hesselbarth and Schaltegger (2014) recommend the curriculum of a higher degree programme to be based on "soft skill training; blended learning, self–directed and collaborative learning, inter-and trans-disciplinarity, and networks for life–long learning with space for informal learning (e.g. Jamieson, 2009)".

Higher education institutions not only play an important role in educating sustainability leaders of the future but also in innovation which is needed for the sustainable and regenerative built environment. For example, the regenerative built environment requires IT intensive materials, which can be improved based on interdisciplinary research and development with the collaboration of professionals in various fields. These researchers need to be trained on sustainability topics in their field so that they can contribute to the research and development process effectively. Matos and Silvestre (2013: 62) emphasized the complexity for sustainable development innovation by saying that:

> *"When compared to traditional market-driven innovation, sustainable development innovation involves additional constraints because social and environmental factors and the preservation of the needs of future generations has to be taken in consideration".*

The innovation process requires collaboration between university and industry as well. Environmental consciousness in industry is important for the success of this process.

6.2.2 Companies and consumers as sustainability leaders

Cities and urban areas are company-and-industry-intensive. Production/manufacturing processes affect the environment especially due to their need for inputs and release of greenhouse gasses. There is need to reduce the adverse effects of the production/manufacturing processes on the cities so that the sustainability and regeneration of cities can be achieved. Companies acting as change agents for sustainability can contribute to the sustainability of cities. Corporate leaders and employees are increasingly recognizing their roles in contributing to sustainability (Lozano, 2012). According to Seuring and Müller (2008), external pressures and incentives by stakeholders influence companies in engaging in sustainability (Matos and Silvestre, 2013). Companies acting as change agents for sustainability can get advantages including the following:

- Corporate sustainability has an economic side (Vallaster and Lindgreen, 2013) as it affects the corporate image and brand. Vallaster and Lindgreen's study (2013) revealed that internal branding relates to the development of a corporate brand and better communicated values, so that the workplace environment is improved to nurture social aspects of a corporation's sustainability. Committed employees feel as if they belong to the organization and take ownership of its fate (Vallaster and Lindgreen, 2013: 298–99);
- Sustainable development can enable employees to fulfil their potential and support companies' business continuity as well as long-term performance (Musson, 2012);
- The companies can have a strong reputation (Goger, 2013);
- The companies can get competitive advantages by adding economic value to both internal and external stakeholders, i.e. lowering production costs through waste reduction (Fiksel *et al.*, 2004 as quoted from Hoejmose *et al.*, 2012).

The main challenges for managers in the companies acting as change agents for sustainability include (Lozano, 2013):

- understanding the structure of their companies, their core competences, and the context of their operations, so that they can choose the combination of initiatives that would address their entire company system, as well as the four sustainability dimensions
- considering that even for the same company, a different set of initiatives might need to be considered in different locations, especially where they have different economic, environmental, social, and legal contexts
- ensuring that the chosen initiative combination creates synergies among them, so that the sustainability dimensions and the company system are fully addressed.

The managers in the companies acting as sustainability leaders need to:

- consider the sustainability principles' implications on the strategic decision-making process of the firm (Schrettle *et al.*, 2014);

- revize current management practices (Schrettle *et al.*, 2014);
- adopt new manufacturing technologies and develop sustainable products (Schrettle *et al.*, 2014);
- collaborate with local governments and firms (Musson, 2012);
- collaborate amongst the supply chain members (Dahan *et al.*, 2010; Perez–Aleman and Sandilands, 2008; Stanton and Burkink, 2008 as quoted in Matos and Silvestre, 2013) and integrate green practices into the supply chain (Schrettle *et al.*, 2014);
- adopt a mix of top–down approach (knowledge and direction is provided by the focal company) and bottom–up approach (practical improvement and learn by doing) (Kaltoft *et al.*, 2007 as quoted in Matos and Silvestre, 2013);
- carry out sustainability strategies with the participation of a diverse number of local stakeholder groups (Matos and Silvestre, 2013);
- collaborate with the stakeholders to increase the chances of finding creative solutions (Matos and Silvestre, 2013);
- encourage pro–environmental behaviour in the workplace (Ones and Dilchert, 2012; Paillé & Boiral, 2013 as quoted in Norton *et al.*, 2014);
- transform the suppliers' mindsets so that they understand the value that they get out of environmental upgrading in terms of differentiation (Goger, 2013);
- reward suppliers for compliance so that suppliers see advantages for environmental upgrading (Goger, 2013).

Managers in the companies acting as sustainability leaders can also influence employees' behaviour in the work environment as well as in the supply chain (Cambra-Fierro *et al.*, 2008). Production and manufacturing is influenced not only by the supply side but also by the demand side, which is affected by the lifestyle and consumption patterns of consumers. Companies, under pressure from stakeholders, increasingly consider issues related to sustainable development (Musson, 2012). For this reason, environment-conscious consumers can lead companies to produce based on sustainability principles enhancing sustainability of the cities. Vergragt *et al.* (2014)'s study revealed that there are a lot of citizens who take responsibility with their lifestyles (i.e. less motorized transport, small but conformable houses, etc.)

6.2.3 Politicians and citizens as sustainability leaders

There is an important mission for the policy makers to act as sustainability leaders both for the welfare of citizens and sustainability of the environment. Furthermore, "there are two-sided relationships between urbanization and economic development. On the one side, it promotes economic development, while on the other side, it is an impediment to economic development of most nations" (Shabu, 2010). Policy makers need to act as sustainability leaders and avoid making populist decisions which might harm the sustainability of cities. For example, they need to support sustainability in city planning and avoid citizens to build slums with the expectations to get their votes (Gökcan, 2014). Such kind of events occur especially in developing countries, causing cities to face challenges such as rapid population increase, lack of economic dynamism, severe infrastructure and service deficiencies, inadequate land administration, poverty, and social breakdown (Rakodi, 2004; Shabu, 2010). This reveals the importance of governance and planning for sustainable cities as described by McCormick *et al.* (2013: 4):

> *"Governance and planning are the key leverage points for transformative change . . . cities and municipalities . . . can be catalysts for change. Sustainable urban transformation requires effective governance and planning [highlighting the critical roles of collaboration and engagement of stakeholders, particularly residents in urban areas".*

Local governments should invest in a sustainable development policy (Musson, 2012). They should also establish new institutions in case they are needed for transformation of the cities into sustainable ones. The need for new institutions has been emphasized by Lockwood (2013)'s study on the forces working for and against the political sustainability of the UK 2008 Climate Change Act. Many politicians should analyse the economic performance of cities and countries not only based on the GDP but also based on the green GDP in which "the damage done to the environment as a whole is factored into the equation to give a clearer picture of the consequences of growing an economy" (WiseGEEK, 2014).

The politicians need to enhance citizens' interest in protecting the environment. For example, as Jordan is facing water scarcity, the former Jordanian Minister of Water Hazem Nasser tried to educate Jordan society through his speeches. He argued that without a sustainable water sector, our economy will not be sustainable. It is absolutely necessary to teach our kids at the elementary level that, we are a water–scarce country (Jordan Business Magazine, 2008; Brown and Crawford, 2009). Providing sustainability-conscious education to young generations can make them rational and lead them not to waste water. Citizens should be integrated into the decision process of the projects. Marschalek's (2008) research on participatory local sustainability projects in seven Chinese villages revealed the effectiveness of integrating the local dwellers into the decision making process in generating ideas for local sustainability-oriented projects and putting the ideas into practice, as well as enabling the project to have a strong bottom–up approach combined with top–down elements. Marschalek (2008) emphasized importance of local projects' visibility for the local dwellers' confidence and their motivation to become engaged in a decision-making process, as the experience of their successful participation in a decision-making process empowers them for self–organization processes or a civil society process.

6.2.4 Media as sustainability leader

The media can be used for many purposes. It is can be used for entertainment, music, sports, reading, watching, computer games etc. Advocacy is used for both economic and social concerns including marketing, advertising, public relations, propaganda, and political communication (Syed, 2014). Media can also act as a change agent through effecting social change by convincing the people (Yahoo Answers, 2014). As media influences the perceptions of the people, it affects their motivation and willingness to act as sustainability leaders. Both mass and social media can have effects on the sustainability of cities. Mass media can act as sustainability leader providing people information on sustainability-related topics such as: the state of the world, climate change, and sustainable buildings and their advantages. The information can be provided in various forms such as documentaries, interviews with specialists, and real–life cases. Social media can act as a sustainability leader enabling people to share videos, pictures, and information. When people have active interactions in social media, their involvement in the design phase of projects can be supported. Citizens' consciousness of sustainability and their involvement in projects as stakeholders can be enhanced through social media. Regenerative construction projects are location-focused and require involvement of the local residents in the design phase so that they provide feedback on local conditions, embrace the project, and take care of the project in the operation phase.

6.3 Conclusions and recommendations

Leadership is crucially important for the establishment and implementation of effective policies needed for the transformation of the cities into sustainable ones. Sustainability leaders can deal with uncertainties and can lead society for change. For the transformation of cities into sustainable ones, there is need for the transformation of individuals in communities into sustainability leaders, increasing societies' awareness and consciousness for sustainability so that they can embrace the change for sustainability and relevant policies. In this way, people will not only be encouraged to participate in the movement but also to lead the change for sustainability. This can accelerate the sustainability transformation needed. Higher education institutions, companies, consumers, politicians, citizens, and media can act as sustainability leaders influencing individuals' perceptions as well as their willingness to act as change agents for sustainability of cities,

- Higher education institutions: Education is the key for environment-conscious future generations, educators and professionals in different disciplines. Sustainability topics need to be integrated into curriculums. Higher education institutions can also carry out innovations needed for the sustainable and regenerative built environment.

- Companies and consumers: Cities are economic hubs. Economic activities affect the sustainability of cities. Production and manufacturing processes should be based on sustainability principles. Consumers' sustainability consciousness, their lifestyles, and consumption patterns affect their expectations and demand from companies, influencing companies' sustainability performance. Businesspeople, policy makers, and society must interact in order to achieve sustainability aims. Such interaction will make them act as change agents for sustainability. Sustainability principles should be integrated into the whole supply chain.
- Politicians and citizens: Politicians should avoid a populist approach. They should support the society to become conscious about the environment and to become more rational, especially through education and regular planning policies. Citizens need to be proactive in policy making and planning. Citizens need to be involved in the decision process of projects.
- Media: Mass and social media influence the perceptions of people, either through providing information or letting people interact. Social media affects individuals' motivation and willingness to act as sustainability leaders. Both mass and social media can be used as training tools for sustainability.

The sustainability leaders (higher education institutions, companies and consumers, politicians and citizens, and media) can influence each other and they are the key ingredients for sustainable cities.

References

Al-Khaddar, R., Wooder, T., Sertyesilisik, B., and Tunstall, A. (2012). "Deep learning approach's effectiveness on sustainability improvement in the UK construction industry", *Management of Environmental Quality: An International Journal*, 23 (2), 126–39.

Bourgeois III, L.J. (1981). "On the measurement of organizational slack", *Academy of Management Review*, 29–39.

Brown, O., and Crawford, A. (2009). "Rising temperatures, rising tensions: Climate change and the risk of violent conflict in the Middle East", International Institute for Sustainable Development (İISD).

Cambra-Fierro, J., Polo-Redondo, Y., and Wilson, A. (2008). "The influence of an organisation's corporate values on employees' personal buying behaviour". *Journal of Business Ethics*, 81 (1), 157–67.

Cortese, A.D. (2003). "The critical role of higher education in creating a sustainable future", *Planning for Higher Education*, 31 (3), 15–22.

Dahan, M.N., Doh, J.P., Oetzel, J., and Yaziji, M. (2010). "Corporate-NGO collaboration: Co-creating new business models for developing markets", *Long Range Planning*, 43, 326–42.

Davis, S.A., Edmister, J.H., Sullivan, K., and West, C.K. (2003). "Educating sustainable societies for the twenty-first century", *International Journal of Sustainability in Higher Education*, 4 (2), 169–79.

Dittmar, M. (2014). "Development towards sustainability: How to judge past and proposed policies?", *Science of the Total Environment*, 472, 282–88.

Fiksel, J., *et al.* (2004). "Environmental excellence: The new supply chain edge", *Supply Chain Management Review*, 8 (5), 50–57.

Goger, A. (2013). "The making of a 'business case' for environmental upgrading: Sri Lanka's eco–factories", *Geoforum*, 47, 73–83.

Gökcan, K. (2014). "Popüler Kültür, Popülizm Ve Politika", available at: http://www.sosyalhizmetuzmani.org/populer_politika.htm (accessed on 20/02/2014).

Guo, F., Chang–Richards, Y., Wilkinson, S., and Li, T.C. (2013). "Effects of project governance structures on the management of risks in major infrastructure projects: A comparative analysis", *International Journal of Project Management*.

Hall, J., and Vredenburg, H. (2003). "The challenges of innovating for sustainable development", *Sloan Management Review*, 45 (1), 61–68.

Hancock, L., and Nuttman, S. (2014). "Engaging higher education institutions in the challenge of sustainability: Sustainable transport as a catalyst for action", *Journal of Cleaner Production*, 62, 62–71.

Hesselbarth, C., and Schaltegger, S. (2014). "Educating change agents for sustainability: Learnings from the first sustainability management master of business administration", *Journal of Cleaner Production*, 62, 24–36.

Hind, P., Wilson, A., and Lenssen, G. (2009). "Developing leaders for sustainable business", *Corporate Governance*, 9 (1), 7–20.

Hoejmose, S.U., and Adrien-Kirby, A.J. (2012). "Socially and environmentally responsible procurement: A literature review and future research agenda of a managerial issue in the 21st century", *Journal of Purchasing & Supply Management*, 18, 232–42.

Jain, S., Aggarwal, P., Sharma, N., and Sharma, P. (2013). "Fostering sustainability through education, research and practice: A case study of TERI University", *Journal of Cleaner Production*, 61, 20–24.

Jamieson, P. (2009). "The serious matter of informal learning", *Planning for Higher Education*, 37 (2), 18–25.

Jordan Business Magazine (2008). "In troubled waters", 11 September 2008.

Kaltoft, R., *et al.* (2007). "Implementing collaborative improvement: Top–down, bottom–up or both?", *International Journal of Technology Management*, 37 (3/4), 306–22.

Linnenluecke, M.K., and Griffiths, A. (2013). "Firms and sustainability: Mapping the intellectual origins and structure of the corporate sustainability field", *Global Environmental Change*, 23, 382–91.

Lockwood, M. (2013). "The political sustainability of climate policy: The case of the UK Climate Change Act", *Global Environmental Change*, 23, 1339–48.

Lozano, R. (2012). "Towards better embedding sustainability into companies' systems: An analysis of voluntary corporate initiatives", *Journal of Cleaner Production*, 25, 14–26.

Lozano, R., Lukman, R., Lozano, F.J., Huisingh, D., and Lambrechts, W. (2013). "Declarations for sustainability in higher education: Becoming better leaders, through addressing the university system", *Journal of Cleaner Production*, 48, 10–19.

Marschalek, I. (2008). "The concept of participatory local sustainability projects in seven Chinese villages", *Journal of Environmental Management*, 87, 226–35.

Matos, S., and Silvestre, B.S. (2013). "Managing stakeholder relations when developing sustainable business models: The case of the Brazilian energy sector", *Journal of Cleaner Production*, 45, 61–73.

McCormick, K., Anderberg, S., Coenen, L., and Neij, L. (2013). "Advancing sustainable urban transformation", *Journal of Cleaner Production*, 50, 1–11.

McIntosh, M., Cacciola, K., Clermont, S., and Keniry, J. (2001). "State of the Campus Environment: A National Report Card on Environmental Performance and Sustainability in Higher Education", Reston, Va.: National Wildlife Federation. http://www.nwf.org/Campus-Ecology/Resources/Reports/State-of-the-Campus-Environment-Report/Read-the-State-of-the-Campus-Environment-Report.aspx [Accessed: 09.09.2014].

Musson, A. (2012). "The build–up of local sustainable development politics: A case study of company leaders in France", *Ecological Economics*, 82, 75–87.

Norton, T.A., Zacherh, H., and Ashkanasy, N.M. (2014). "Organisational sustainability policies and employee green behaviour: The mediating role of work climate perceptions", *Journal of Environmental Psychology*, 38, 49-54.

O'Brien, W. and Sarkis, J. (2014). "The potential of community–based sustainability projects for deep learning initiatives", *Journal of Cleaner Production*, 62, 48–61.

Perez–Aleman, P., and Sandilands, M. (2008). "Building value at the top and at the bottom of the global supply chain: MNC–NGO partnerships", *California Management Review*, 51 (1), 24–49.

Posch, P. (1999). "The Ecologisation of schools and its implications for educational policy", *Cambridge Journal of Education*, 29 (3), 341–48.

Radywyl, N., and Biggs, C. (2013). "Reclaiming the commons for urban transformation", *Journal of Cleaner Production*, 50, 159–70.

Raivio, K. (2011). "Sustainability as an educational agenda", *Journal of Cleaner Production* 19 (16), 1906-07.

Rakodi, C. (2004). "African towns and cities: Power houses of economic development or slums of despair?", Paper at "City Future" Conference, University of Illinois, Chicago.

Rands, G. (2009). "A principle–attribute matrix for environmentally sustainable management education and its application: The case for change–oriented service-learning projects", *Journal of Management Education*, 33 (3), 296–323.

Rangarajan, K., Long, S., Tobias, A., and Keister, M. (2013). "The role of stakeholder engagement in the development of sustainable rail infrastructure systems", *Research in Transportation Business & Management*, 7, 106–113.

Radywyl, N., and Biggs, C. (2013). Reclaiming the commons for urban transformation. Journal of Cleaner Production 50, 159–170.

Shabu, T. (2010). "The relationship between urbanization and economic development in developing countries", *International Journal of Economic Development Research and Investment*, 1 (2/3).

Schrettle, S., Hinz, A., Scherrer-Rathje, M., and Friedli, T. (2014). "Turning sustainability into action: Explaining firms' sustainability efforts and their impact on firm performance", *International Journal of Production Economics*, 147, 73–84.

Schmitt, K. (2011). "Going big with big matters: The key points approach to sustainable consumption", *GAIA – Ecological Perspectives for Science and Society* 21 (2), 91–94.

Seuring, S., and Müller, M. (2008). "From a literature review to a conceptual framework for sustainable supply chain management", *Journal of Cleaner Production*, 16 (15), 1699–1710.

Shephard, K. (2008). "Higher education for sustainability: Seeking affective learning outcomes", *International Journal of Sustainability in Higher Education*, 9 (1), 87–98.

Stanton, J.V., and Burkink, T. J. (2008). "Improving small farmer participation in export marketing channels: Perceptions of US fresh produce importers", *Supply Chain Management*, 13, 199–210.

Stubbs, W., and Schapper, J. (2011). "Two approaches to curriculum development for educating for sustainability and CSR", *International Journal of Sustainability in Higher Education*, 12 (3), 259–68.

Syed, S. (2014). "Media As a Change Agent, Mass Media", http://saeedasyed.blogspot.com.tr/p/mass–media.html [Accessed: 28.03.2014]

Tukker, A. (2013). "Knowledge collaboration and learning by aligning global sustainability programs: Reflections in the context of Rio+20", *Journal of Cleaner Production*, 48, 272–79.

United Nations (2014). "The future we want: Sustainable cities", http://www.un.org/en/sustainablefuture/cities.shtml#facts [Accessed: 28.03.2014].

UN–Habitat (2008). *State of the World's Cities 2008/2009 Harmonious Cities*. Earthscan, London.

Vallaster, C., and Lindgreen, A. (2013). "The role of social interactions in building internal corporate brands: Implications for sustainability", *Journal of World Business*, 48, 297–310.

Vergragt, P., Akenji, L., and Dewick, P. (2014). "Sustainable production, consumption, and livelihoods: Global and regional research perspectives", *Journal of Cleaner Production*, 63, 1–12.

Waddock, S. (2007). "Leadership integrity in a fractured knowledge world", *Academy of Management Learning and Education* 6 (4), 543–57.

WEC and Net Impact (2011). "Business skills for a changing world: An assessment of what global companies need from business schools", World Environment Center & Net Impact, available at: http://www.greenbiz.com/sites/default/files/Net%20Impact_WEC%20Report_FINAL.pdf.

WiseGEEK (2014). "What Is Green GDP?", http://www.wisegeek.com/what–is–green–gdp.htm.

Yahoo Answers (2014). "How is mass media an agent of social change?", https://answers.yahoo.com/question/index?qid=20081021001501AArMDyb.

Part III

Leadership for sustainable built environment

7 The role of organizational leadership in the delivery of sustainable construction project practices

Alex Opoku and Chris Fortune

7.1 Introduction

Business organizations in the construction industry need leadership that provides a collective vision, strategy and direction to enable them to help deliver society's common goal of a sustainable future. The role of leadership in improving performance and innovation in the construction industry has been receiving increasing attention in recent times. However, less attention has been given to the capability of organizational leadership in construction organizations to promote the delivery of sustainable practices in their construction projects. The change towards the adoption of more widespread sustainability practices in the delivery of construction projects is a process that presents a challenge to organizational leadership. This chapter therefore considers the role of organizational leadership within United Kingdom (UK) business organizations involved with the delivery of sustainability practices in construction projects. The principal types of business organizations involved in the delivery of construction projects can be classified as being either contractor or consultant types of organization. Initially, relevant terms are defined and the context for leadership and sustainable construction practices is established before the consideration of the roles that leadership could play in both contractor and consultant types of organizations that are involved in the delivery of sustainable construction projects. The chapter helps to make the case for further work to be undertaken to explore the link between organizational leadership and the adoption of sustainable construction project practices.

7.2 Definition of terms

The role of a leader can be formally or informally developed within a group and it is the role of the leader to organize, motivate and assign tasks in the group so as to enable it to achieve its goals. Taylor (2008) believes that anyone at any level in an organization could potentially be considered as a leader at any given point in time if he or she is involved in a process of influence that involves encouraging and influencing others to adopt sustainable practices. Similarly, Gattiker and Carter (2010) indicate that organizational leadership seeks to influence other individuals (subordinates, superiors, and / or peers) within their organizations to achieve specific aims and objectives. Leadership is therefore defined for the purposes of this work as a process of influencing a group of individuals within an organization to accomplish a common goal that is related to construction project sustainability practices while not necessarily being in an executive position. Sustainability practices in this chapter refer to processes and activities involved in achieving the sustainable delivery of construction projects. These practices are at pre–construction, construction and post construction stages of a project's life cycle and may include sustainable design, procurement, site waste management, materials and resources use, whole life costing etc.

7.3 Leadership for sustainability

Leadership is considered as a process of influencing organizational direction and vision that occurs through the relationships between leaders and their followers (Taylor *et al.*, 2011). Leadership is generally agreed to be an important factor in achieving business success in any organization. Jing and Avery (2008) argue that although the concept of leadership still lacks clarity and agreement in literature it has a very significant influence on organizational activities including the delivery of organizational sustainability. The study and the understanding of leadership and its relationship to sustainability is still in its early stages and requires more work to establish the important role of organizational leaders in implementing sustainable strategies. For instance, Chan and Cooper (2007) conducted research through in–depth interviews with fifteen leaders of the UK construction industry. The work revealed that the understanding of construction leadership for sustainability could be said to be primitive when compared with the rather more mature understanding of developments in mainstream leadership research. The significance of organizational leadership for sustainability is increasing rapidly as a result of increasing amounts of national and international legislation related to sustainability and the need for organizations to innovate continuously to meet their ever–changing business environments.

As a result organizational leadership needs to take bold steps to move beyond efficiency, legislative compliance or just being seen to be green. Sustainability is now viewed by many organizations as being part of a strategy for long-term business survival and success (McCann and Holt, 2010). Similarly, the link between organizational leadership and sustainable practices is now seen as being vital; for instance, Parkin (2010, p. 89) asserts, "Leadership is a vital ingredient for achieving sustainability. Without it sustainability will never make it in government, business or anywhere". Organizations are now being encouraged to fundamentally change the way they operate from focussing on the short–term maximization of value for their shareholders to now paying attention to the economic, social and environmental effects of their operations. Leaders have a significant role to play in construction industry–related organizations as the ability of such organizations, irrespective of their level of maturity, to pursue the sustainability agenda is influenced by the commitment and conviction of their leadership's approach towards sustainability. Leaders need to communicate the importance of sustainability and establish a culture of integrating sustainability into the day-to-day management decisions of their organizations (Avery, 2005). In order for this to be achieved it is asserted that an organization's leadership needs to be developed that is self–aware and committed to sustainability practices. Once this has become established in the leadership of an organization then such practices will be supported by other colleagues within that organization. As a result the organization as a whole can then direct the actions of its project-related stakeholders towards the achievement of common project-related goals of sustainable construction project practices.

7.4 Sustainable construction project practices

Sustainable construction project practices seek to achieve a balance between the project–related pressures of environmental resource protection, social progress and economic growth both now and for the future. It has been found that some construction organizations that have adopted sustainability practices subsequently report greater levels of profits as well as improved productivity and enhanced health and safety management records. Similarly, such approaches also promote wellbeing through the provision of a better workplace for staff and the generation of community benefits in the locations in which their projects are executed (SECBE, 2005). It is accepted as common practice now for construction organizations to produce building designs that minimize energy usage and water resources, minimize waste and prevent pollution as well as preserving and enhancing the local ecological biodiversity in which their projects are located. Singh (2007) argues that in addition to such practices the industry is also adopting appropriate practices in terms of materials re–cycling, materials transportation and the choice of methods used for the actual construction of the project. It is widely accepted that all such factors can have negative impacts on the environment in terms of energy use and the generation of high levels of noise, waste and dust as well as their potential impact on the quality of air, water and soil. Such environmentally friendly practices are well established and will therefore minimize the overall environmental impact of the built assets generated and maintained by the construction industry but the real challenge is now to adopt leadership measures that will promote all aspects of the triple bottom line so as to make organizations adopt a truly sustainable, not just environmentally friendly, approach to construction project practices.

7.5 Organizational leadership roles

Sustainability is better implemented in an organization when there is an active leadership within the company that is willing to champion the sustainability approach. Organizational leaders play vital roles in the promotion of sustainability practices in construction organizations. Opoku (2012) undertook a doctoral study of leaders and their roles in relation to the sustainability agenda in construction consultant and contractor organizations. That work adopted a qualitative research approach to the collection and analysis of data from a series of semi–structured interviews with such organizational leaders. As a result it was found that construction organizations require committed organizational leadership to champion the sustainability agenda through the day–to–day activities of the company. Organizational leaders provide vision and are responsible for implementing sustainability strategies throughout the company and also promoting the organization's sustainability policies to external companies and clients.

Sustainability professionals in construction organizations work with clients to advise them on how best to embed sustainability into project briefs. A unique role of organizational leaders from contractor types of organizations is that they can become involved with lobbying government for legislation to achieve sustainable change in the UK construction industry. It was noted that organizational leaders responsible for sustainability have a duty to challenge construction processes within the organization to ensure that the organization adopts practices that link sustainability with the overall company's vision and strategy on sustainability (Opoku, 2012). In addition, organizational leaders with sustainability roles can become involved with leading, encouraging, mentoring and supporting all other employees, departments and units towards the implementation of sustainable construction practices. Beyond the above roles, such leaders are also charged with the development of sustainability guidance notes and policies for their organizations (Opoku, 2012).

Organizational leaders have a role in helping to promote sustainable construction practices by providing training and awareness courses to staff on sustainability issues, and in so doing driving forward the sustainability agenda, disseminating sustainable construction best practice and monitoring project related sustainability targets and performance. Despite the potential benefits of current sustainability policies to drive improvements in building performance, it has not changed the mind-set of many construction practitioners who look at project-related sustainability practices as an add-on target which they are forced by legislation to satisfy. The challenge of changing the culture and behaviour of such employees and other project stakeholders towards the ready acceptance and adoption of sustainability practices has been identified as one of the key leadership roles (Opoku and Fortune, 2011). Other organizational leadership roles could include the promotion, use and maintenance of an organization's environmental management system and its project appraisal models to suit local, national and international government policy requirements.

As a result most organizational leaders act as "sustainability integrators" who are responsible for the integration of sustainability strategies internally within the various business units of an organization. This role would involve coordinating all the different departments of an organization and influencing how business is approached by ensuring sustainability practices are at the heart of all an organization's business activities. All the key organizational leadership roles that have been found to be needed to promote sustainability practices in both contractor and consultant types of construction organizations have been summarized and presented in Table 7.1.

The nature of the exploratory work undertaken by Opoku (2012) was able to identify the key role of the organizational leadership in the adoption of sustainability project related practices.

The findings showed that one of the most important roles of leadership in promoting sustainability is to "formulate policies, implement procedures and disseminate best practice", followed by "helping to develop sustainability strategies" for their organizations, with the role to "lead the drive to make the organization sustainable" as the third most important. Figure 7.1 shows the organizational leadership roles relevant to the consultant and contractor types of construction organizations.

There are both common and diverse roles performed by organizational leaders charged with the promotion of sustainability practices in contractor and consultant organizations. A unique role cited performed by leaders from the contractor organizations is the lobbying of government for sustainable policy changes in the UK construction industry. Further work is, however, needed to fully ground these findings so that an emergent framework or

Table 7.1 Organizational leadership roles in promoting sustainability practices

Leadership and the Promotion of Sustainability in Construction Organizations
Providing direction, setting the vision and championing the drive to make the organization sustainable
Assisting the rest of the company on sustainability issues by helping, encouraging and mentoring staff
Devising strategies and procedures on implementation for the organization to disseminate sustainability best practice
Driving forward the sustainability agenda within the organization by providing guidance notes, training and awareness courses for staff
Marketing and promoting sustainability by talking to potential clients and other project-related stakeholders about the benefits of sustainability
Lobbying government for legislation to promote sustainable change
Acting as sustainability integrator across all sections and departments of the organization
Developing, using and maintaining an environmental management system across the organization
The promotion of a culture of change from compliance with sustainability-related legislation to one of organizational pro-activity
Setting and monitoring sustainability targets and performance

Source: Adapted from Opoku (2012)

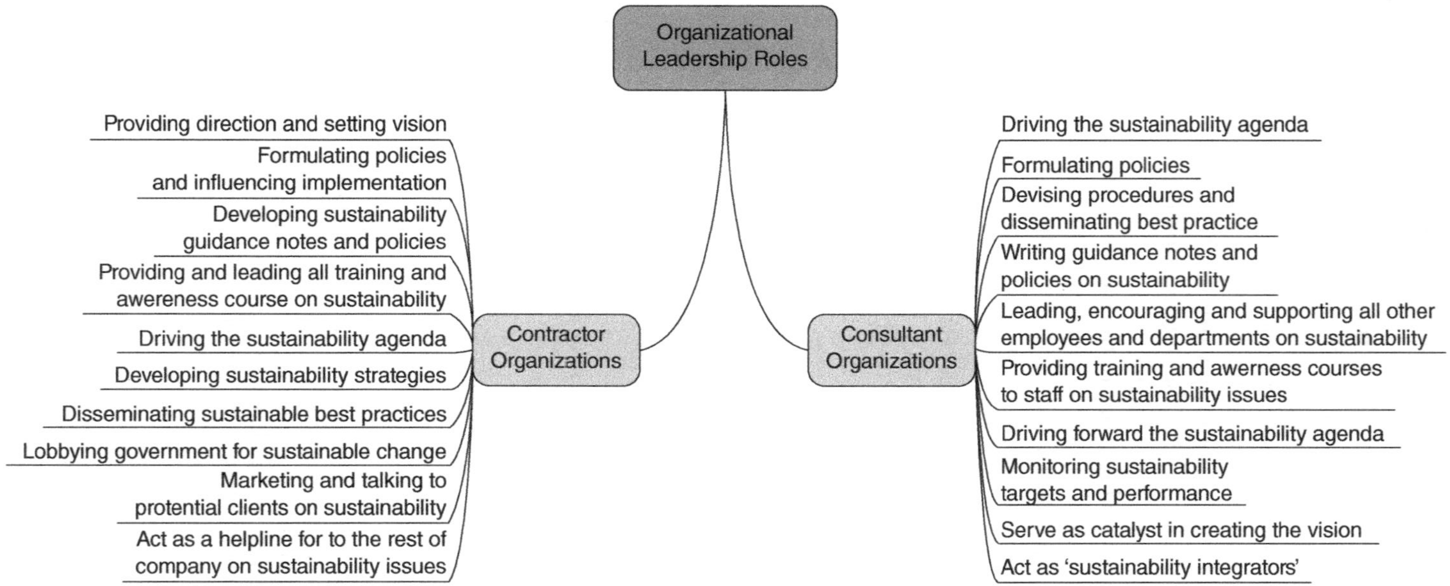

Figure 7.1 Type of construction organization and leadership roles in promoting sustainability

decision aid could be developed to assist practitioners seeking to promote sustainability practices in their construction industry organizations.

7.6 Summary

In the UK, leadership roles in construction industry organizations can differ depending on the size and type of organization especially in relation to the promotion of sustainability practices. Sustainability champions can be found within such organizations at all levels. This chapter addressed the gap in construction management literature on the link between organizational leadership and sustainability in construction. The chapter provided an insight into the potential roles of organizational leaders in the promotion of sustainability practices. The potential roles that were identified following an exploratory study in the field were summarized in Table 7.1. The leadership roles listed in Table 7.1 could also serve as guide to other organizational leaders seeking to promote sustainable construction project delivery practices. The findings of the study described provide the empirical evidence to support the assertion that organizational leadership has a key role to play in the adoption of sustainability practices. The chapter helps to make the case for further work to be undertaken to explore the link between organizational type, leadership approaches and the adoption of sustainable construction project practices.

References

Avery, G. (2005), *Leadership for sustainable futures: Achieving success in a competitive world*, Edward Elgar Publishing, Cheltenham.

Chan, P. and Cooper, R. (2007), "What makes a leader in construction? An analysis of leaders in the UK construction industry", In: *CIB World Building Congress*, 13–17 May 2007, Cape Town, South Africa, pp. 498–510.

Gattiker, T. F. and Carter, C. R. (2010), "Understanding project champions' ability to gain organizational commitment for environmental projects", *Journal of Operations Management*, Vol. 28 No. 1, pp. 72–85.

Jing, F. F. and Avery, G. C. (2008), "Missing links in understanding the relationship between leadership and organizational performance", *International Business and Economics Research Journal,* Vol.7 No. 5, pp. 67–78.

McCann, J. T. and Holt, R. A. (2010), "Servant and sustainable leadership: Analysis in the manufacturing environment", *International Journal of Management Practice*, Vol.4 No. 2, pp.134–48.

Opoku, A. (2012), "Promoting sustainability practices through leadership within UK construction organizations", (Unpublished Doctoral Thesis), University of Salford, Manchester–Salford.

Opoku, A and Fortune, C. (2011), "The implementation of sustainable practices through leadership in construction organizations", In: Egbu, C. and Lou, E.C.W. (Eds.) *Procs 27th Annual ARCOM Conference,* 5–7 September 2011, Bristol, UK, Association of Researchers in Construction Management, 1145–54.

Parkin, S. (2010), *The positive deviant: Sustainability leadership in a perverse world*, Earthscan, London.

SECBE (2005), *An introductory guide to best practice in construction*, South East Centre for the Built Environment, Reading.

Singh, T. P. (2007), "Sustainable Construction", *Sustainability Tomorrow*, CII-ITC Centre for Excellence in Sustainable Development, Delhi.

Taylor, A., Cocklin, C., Brown, R. and Wilson–Evered, E. (2011), "An investigation of champion–driven leadership processes", *The Leadership Quarterly*, Vol. 22 No. 2, pp. 412–33.

Taylor, A. (2008), "Promoting sustainable practices: The importance of building leadership capacity", *Proceedings of the Enviro 08 Conference*, 5–7 May, Melbourne, Victoria.

Part IV

International perspective and case studies

8 Challenges to leaders in promoting innovative and sustainable social housing in the UK

Yamuna Kaluarachchi

8.1 Introduction

The UK government has set a challenging 80 percent reduction target in Carbon Dioxide (CO_2) emissions by 2050. The residential sector accounts for around 30 percent of the total final energy use and produces 157.2 $MTCO_2e$ per year, which accounts for 27 percent of the total UK CO_2 level by end-user sector (DECC, 2012). According to the Department of Communities and Local Government, the average English home produces 5.8 TCO_2 per year (DCLG, 2012). Notwithstanding the government's plan to increase the UK housing stock by 240,000 homes a year, approximately 80 percent of today's dwellings will still be standing in 2050 (Boardman, 2008). Improving the energy efficiency of UK existing housing stock has been made a priority within the UK government's Energy White Papers (DTI, 2003 & 2007) as an effective, clean, safe and cost effective approach to meet the carbon reduction targets (Gaterell *et al.*, 2005, Power, 2006). However, the uptake of effective energy efficient strategies within the UK is currently not sufficient to achieve the necessary CO_2 reduction targets (DECC, 2009, SDC, 2005). Whilst it is generally accepted that retrofitting low carbon solutions to existing buildings is more complicated than installing them in 'New Build', these complications are largely socio-economic rather than technical (Boardman, 2007; WWF, 2008). In the social housing sector, the issues faced are not only technical and socio-economic but also organisational and managerial and require innovation in thinking and technology to face the challenges of achieving sustainable housing. Leaders and Senior Managers have to assess risks and priorities actions to meet the demand as well as achieve value for money. Thus, understanding the barriers in promoting innovative sustainable housing and identifying suitable drivers that can overcome them is essential if the UK housing sector is to have any chance of reducing its CO_2 emissions to meet government targets.

Innovation in the current context has to incorporate issues of social, environmental and economic sustainability and organisations, their structure and leadership have a major influence in this process. Developing a culture of innovation in organisations and industry appears to be vital in triggering innovation. The main driving forces are the ideas of stakeholders: customers, management, marketing personnel and production personnel, as they focus on problem fixing and developing new ideas. The lack of proper qualifications, training, and access to cutting edge knowledge and technology; fear of taking risks; and the culture and mind-sets of the particular organisation could all be contributing factors. For an organisation to be competitive, specialist skills, consultancy services and professional expertise are needed. Forward thinking, considering whole life values rather than short-term demands, profits and balancing books are essential in this process. This new way of thinking can be driven by leaders and stimulated by research and development within the organisation or externally; exposure to innovative technology and projects; promoting best practice; and specialist training. While some European countries, especially Scandinavian, move forward by bringing their sustainability agenda to the public sector, up until recently the UK seems to have been slow in embracing these ideas. A number of reviews of the construction industry provided waves of re-structuring and re-inventing, but long-term sustainability in improved products and processes for better performance, efficiency and mainstreaming of innovative application of renewable and low carbon technology serving the built environment is yet to come. While funding remains a major constraint, there are many other issues that directly or indirectly influence this process (Kaluarachchi *et al.*, 2007a). The combination of attitudes towards risk and a wariness of innovative solutions result in organisational barriers to the wider uptake of low carbon technologies. Creative leadership is essential in overcoming these barriers. Another key factor in this process is the 'innovation capability of an organisation', which locates and develops potential innovations that can be transferred into the mainstream (Lawson *et al.*, 2001).

This chapter presents the results from a number of research studies carried out in the social housing sector over a decade (2000–2010) to highlight key issues that are important in organisational and managerial terms in organisations in implementing innovative technologies in sustainable refurbishment. The research methodology was based on case study monitoring, action research, consultation workshops and in-depth survey of related stakeholders. The aim was to identify barriers and drivers that influence organisations and highlight the challenges leaders face in achieving sustainability targets. The chapter will also present the findings of a research study into the level of perceived organisational sustainability, attitudes to risk and innovation amongst social housing providers. A detailed survey was carried out to establish the level of sustainability that was achieved and attitudes towards risk of the responding organisation along with the experiences of sustainable refurbishment of their housing stock.

8.2 Innovation and organisations

Many organisations that have introduced a unique product or service often fail to develop effective long-term strategies to sustain them. Problems in commercialising and mainstreaming new products often arise from poor marketing or organisational structures, rather than a lack of usable technology. An ability to manage innovation as part of the overall corporate strategy is as important as the innovation itself (Barlow, 1999). This ability can be affected by previous experiences, fear of change, impact of innovation on existing organisational hierarchies, work processes or management structures. Investors concerned about risky investment in new techniques or products can also hinder this ability. Managing innovation essentially involves mediating between external forces for change and internal forces for stability and giving leadership in the transition.

Strategies that organisations have developed to stimulate innovation in social housing mainly relate to process, procurement, learning, benchmarking and training issues. Factors of influencing the market, peoples' pre-conceptions, planning and building regulations and finance can also be identified to a lesser degree. Several developments in Building Process Innovations offer the potential to reduce costs. Innovation in materials technology has influenced a wide range of house building products. These include improved plastics, new ceramic technologies, and high thermal and acoustic performance and partitions, which allow a wider choice of finishes (Pan *et al.*, 2005). Some house builders have taken a mixed approach of taking advantage of both off-site and on-site production. The mixed approach favours incremental, rather than radical, innovations. Hooper and Nicol (2000) highlight the dominant practice among most large house building companies of continual incremental modifications to existing standard house types, rather than the creation of new designs. Seaden *et al.* (2003) studied two sets of variables of business environment and business strategies on innovative practices and suggest that innovation leads to improved competitive advantage and greater profitability. However, innovation is risky, requires significant investments and is often resisted within the firm. This justifies the strategies developed by house builders on learning, benchmarking and training.

8.2.1 Organisational capability to innovate

Cohen and Levinthal (1990) argue that industrial research and development (R&D) not only generates new information but also improves the ability of firms to absorb knowledge developed outside the firm. The type of staff employed provides an indication of a firm's capability to develop, manage and utilise new technical knowledge. Other issues equally important are organisational structure and culture; the nature of internal and external communications; coordination and feedback mechanisms; the ability to codify knowledge; and the type and use of information and communications technologies (Cohendet *et al.*, 2000). Companies associated with fast-moving science and technology sectors usually invest more intensively in R&D than most construction organisations. By other industries' standards investment by government and construction firms in R&D is very low, particularly in the UK; this is not the case in some other countries such as France, Japan or Scandinavia (Gann, 2001). It follows, from the argument that lack of internal R&D capability in construction indicates that many firms are unlikely to have the capability to absorb the results of academic research or work published in journal articles (Cohendet *et al.*, 2000). Technological progress across the sector is therefore likely to be slow. When faced with the prospects of technological change, the majority of construction firms are recipients of innovation first exploited in other sectors, or by a few construction market leaders. Even when a firm has the technical competence to absorb new ideas, it may not have the internal structure, systems and cultural attributes necessary to capitalise on research results.

8.2.2 Organisational structure and communication

Organisations and their structure have a major influence in the innovation process. Visionaries who have corporate influence can drive innovation and influence the market growth, but will need support from other organisations in stabilising the process and creating the demand that is needed to establish the market. The monitoring of Amphion Consortium of Register Social Landlords (RSLs) (2001–2004) to establish high quality housing designed and procured in line with a new procurement agreement illustrate how organisational structure and communication have a marked influence on innovation projects. In this arrangement a strategic partnering contract was set up with a single contractor who developed an award-winning modern pre-fabricated timber frame housing system. The RSLs agreed to procure 2000 new house units over a four-year period and the research project exploited the opportunity to study a major innovation programme and identify what key lessons could be learnt. The main aim of the research was to set, monitor and compare the Key Performance Indicators (KPIs) and map the cause and effect relationships within the change programme. The research methodology was based on case study monitoring and action research and a range of questionnaire surveys, detailed interviews with key project personnel, examination of site meeting notes and general feedback reviews were undertaken to identify good and bad practices associated with each project (Kaluarachchi *et al.*, 2007b).

The innovative timber frame system needed more research and development input prior to implementation to reduce defects and the associated costs. The components were over-designed to minimise risk which made the product expensive compared to other timber frame housing systems in the market. These increased costs and the defects resulted in lack of trust in the contractor and the volume of demand initially forecasted never materialised. As a result the contractor was the subject of several takeover bids by rivals and experienced a number of problems with both the supply of the timber frame housing

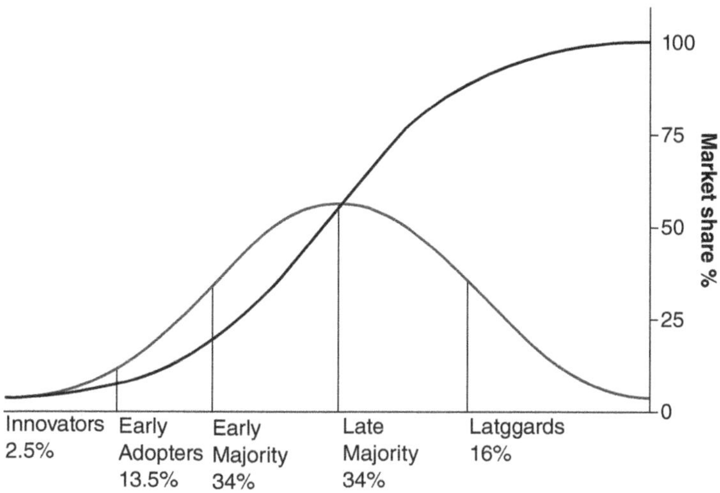

Figure 8.1 The 'Diffusion of Innovations' Curve

system and site personnel, which compromised the quality of construction and resulted in a high turnover of site-based operatives. Key lessons learnt from the initiative:

- Organisations need to be fully committed to the innovation programme.
- Risk management processes needed to be evaluated, agreed upon and in place prior to commencing the project.
- Innovative processes require a change in mind-set at all levels within the organisations. Effective mechanisms should be put in place to ensure that everyone understands the joint goals and knows their part in the overall process. Better communication at all levels in delivering product is essential.
- Training was identified as an essential ingredient in this process. The lack of familiarity with the innovative approach illustrated the need for formal training for all project managers, prior to commencement of new projects. There was also the need for support systems in terms of knowledge and information to be in place for frontline staff.
- Communication and co-ordination, which lead to continuous improvement of services and products, emerged as some of the key drivers of the process.
- Even though the government encourages initiatives, such as that monitored in the research study, there is little flexibility in support systems to assist in sustaining them.
- Continuous improvement process that would feed information from the site and different stakeholders who were involved with the project, was never implemented.

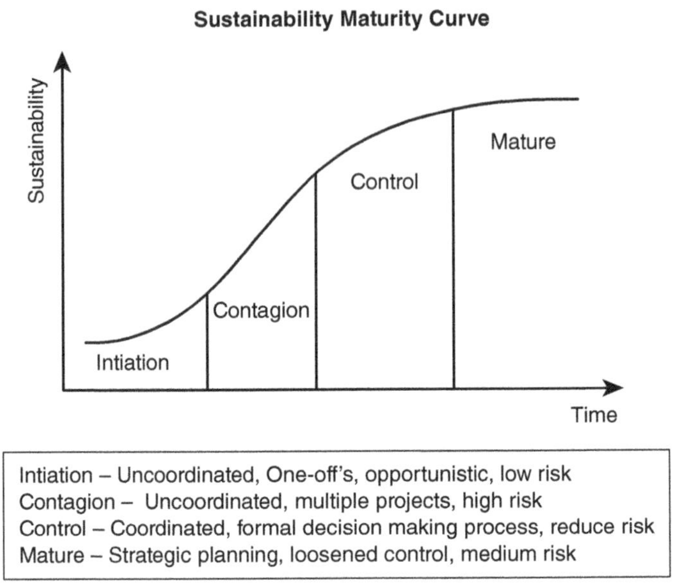

Figure 8.2 Sustainability Maturity Curve

8.3 Uptake of innovative environment technologies in the social housing sector

The EPSRC SUE-IDCOP programme (2005-2008) provided knowledge to underpin the improved sustainability of existing buildings. The overarching aim was to find ways to improve the performance of existing building envelopes, which reduce the consumption of non-renewable resources over the whole building life cycle in a way that is economically viable and socially acceptable. There are many innovative environment technologies that are readily available in the market for this purpose. Case study examples here in UK and extensively throughout Europe show that these technologies can be used effectively and economically in new build housing. In UK, there is very little evidence of their use in routine maintenance and refurbishment. The aim was to identify and review the barriers that stakeholders face in promoting innovative environment technologies in social housing refurbishment. A number of stakeholder consultations were undertaken in relation to three sectors in the procurement of social housing: management, development and the maintenance sectors. It was considered under the following criteria: energy performance, water performance, waste management, durability and flexibility (whole life performance), health and wellbeing of tenants (quality of life issues).

The results illustrated that:

- The capital costs of most of these technologies are significantly higher than the available budgets and the potential cost savings in utility bills. Also, the tangible benefits of employing renewable technologies are usually long-term and do not result in quick savings. Value for money is a major governing factor in the decision-making process and for the technology to be employed, benefits should outweigh the costs incurred.
- The technology should be proven and fully demonstrated prior implementation. Confidence levels in the new products are low due to high costs in demonstration projects (example – Amphion project) and occupants and organisations are reluctant to take the risks.
- There are quite a lot of products and systems in the market but very little information about their long-term performance, durability and ways in which they can directly reduce cost. More information about whole life performance and cost savings is needed and should be made available to the RSLs.

All the above factors illustrate that there is a demand for more information, effective communication and research and development. Informing and educating tenants and organisations about the long term benefits and whole life cost value seem crucial in implementing innovative technologies. Research and development is essential to bring the cost down and increase market potential.

8.4 Characteristics of innovators and innovations

Rogers (2003) characterises innovators in his theory of 'diffusion of innovations' (Figure 8.1), where the ideas are not instantly adopted, but instead there are differing rates of adoption of an idea by different participants in a market or industry. There are the 'innovators', who find and implement new technologies at their inception and adapt and use these ideas to stay ahead. The 'early adoptors' recognise and adopt the new technology or innovations ahead of the mainstream, and adopt it early to gain a competitive advantage. There are the 'early majority', who recognise that a shift is happening in their industry and adopt the change at the same time as others do or slightly before. There are the 'late majority', who recognise that a shift has occurred in their industry, and make the changes required to adopt the new technology and stay in the mainstream. Then there are the 'laggards', a group who are slow to recognise and adopt the new technology and its implications

Rogers (2003) examined the factors that governed the uptake of innovation, identifying the degree of relative advantage; compatibility with existing values and practices; simplicity and ease of use; trialability; and observable results as key drivers that influence the speed of uptake of innovative solutions. Earl (1989) applied a similar approach to Rogers' when he investigated the uptake of Information Technology in data processing organisations and identified a multiple 'S' curve model to describe uptake of the then-innovative technology, attributing positions along the 'S' curve to stages of organisational maturity (Initiation, Contagion, Control and Maturity). By observing organisational characteristics at each stage of the S curve Earl identified operational and strategic policies and management orientations that governed the degree to which IT had effectively been integrated into the business. Hinks *et al.* (2007) applied the 'S' curve model to facilities management, using it to distinguish between sustaining and disruptive innovation. Hinks argued that if innovation follows a continuous 'S' curve (sustaining innovation), then it can only produce innovation that is incremental on what has gone before, as illustrated in Figure 8.2. As an innovation reaches the end of the 'S' curve journey, then the ability of the next incremental improvement to deliver meaningful advantage will diminish and as such its uptake rate will decline. These theories were combined in this project to investigate the uptake of sustainable technologies in the refurbishment of UK social housing.

8.5 Organisational structure and perceived sustainability

A detailed questionnaire survey was carried out (2009) with a sample of Housing Associations and local authorities to survey the innovative environment technologies and processes that are implemented in the sustainable refurbishment of UK social housing. The questionnaire, sent to the senior managers and decision makers, examined the barriers and drivers to sustainable refurbishment projects and related these to organisational characteristics and management attributes. The questionnaire also sought to measure the level of perceived sustainability by the organisation itself according to four levels of sustainability. The questionnaire comprised 15 questions covering: interpretation of the sustainability agenda; formal policies and business procedures; perceived drivers and barriers to sustainable refurbishment; and the decision making process / business case for action. A total of 500 questions were distributed and 57 responses we received representing the response rate of 11.4 percent. Even though the response rate was low, the study had the assurance that all respondents were Chief Executives or Senior Managers who would prioritise and make the strategic decisions of organisations.

Respondents were asked a series of questions about their understanding of sustainability and the relative importance that they believed each attribute should contribute towards a sustainability assessment. All respondents identified that sustainability was about balancing environmental, social and economic performance of their housing stock but the relative importance that they attached to each attribute varied depending on where they placed themselves on the organisation's and national sustainability agenda. The respondents were asked to rate the sustainable refurbishment actions carried out by their organisation according to four levels.

Level 1 - (Low) Actions are uncoordinated one offs, low risk and opportunistic (Initiation)

Level 2 - (Low/Medium) Actions are uncoordinated, in multiple projects, high risk (Contagion)

Level 3 - (Medium) Actions are coordinated, supported by formal decision-making process to reduce risk (Control)

Level 4 - (High) Actions are strategically planned, supported by formal decision-making process and at medium risk (Mature). Of the 57 respondents, 17 placed themselves at level 1, 20 at level 2, 14 at level 3 and 2 at level 4 (Figure 8.3).

Respondents were presented with potential drivers relevant to their refurbishment decisions and asked to rank these in priority order. The most important drivers were: tenant satisfaction (20.5 percent); government policy (20 percent); available funding/business support (13.9 percent); legislative support (9.4 percent); and education/knowledge (9.4 percent). With regards to the barriers for sustainable refurbishment: lack of funding [17.7 percent]; high initial capital cost [17.3 percent]; long payback periods [13.8 percent]; value for money [12.1 percent]; fear of risk [8.3 percent]; and lack of knowledge [8.3 percent] were perceived as major barriers. Respondents were also asked to identify the governing factors that determined the level of sustainable refurbishment that they believed was required. Eighty percent of respondents identified the state of their housing stock followed by organisational leadership (56 percent), return on investment (56 percent), tenant buy-in (52 percent), and confidence in the solution (52 percent) as the most important factors when identifying which sustainable refurbishment project to undertake.

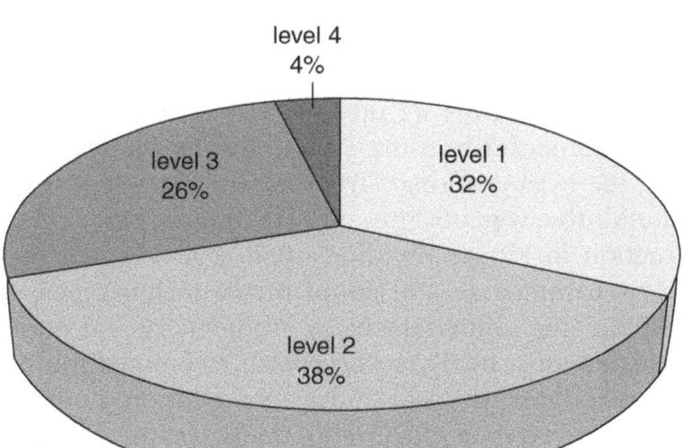

Figure 8.3 Perceived sustainability by organisations

Table 8.1 Organisations' decision-making process for innovative sustainable refurbishment

Stage	*Decision-making process*
Level 1 (Initiation)	Organisations have no executive manager responsible for overall delivery of sustainable refurbishment. The actions are one offs.
Level 2 (Contagion)	Overall responsibility located at the executive level: a short development time between board level decisions to local implementation: decisions are based on the outcomes of a stock condition process: a long-term plan and vision for their housing.
Level 3 (Control)	Organisations had developed specific procurement routes for innovative technologies; begun to involve tenants in decision-making; a semi structured options appraisal process in place to distinguish priorities.
Level 4 (Mature)	A senior-level manager is responsible for the overall delivery of the sustainability activities. High levels of tenant involvement throughout the process; individual project monitoring against a long-term asset management plan.

An attempt was made to identify differences in the decision-making hierarchy or implementation routes between organisations at different stages of perceived sustainability levels as in Figure 8.3. Whilst many management and decision-making attributes are common to all respondents, the degree to which they influence sustainable refurbishment decisions vary. The results illustrated that in Levels 1, 2 and 4 decision-making is devolved and reporting / monitoring is flexible. In Level 3 management approaches were more prescriptive and reporting was more formal (Table 8.1).

8.6 Discussion

Case Study 1, Amphion (2001) was set up by a visionary who wanted to achieve real change in the social housing sector. But as the lessons from this experience show, the organisational structure and the support needed to facilitate such an initiative was not present. The report 'Offsite Modern Methods of Construction in House Building' (Pan, 2005) stated that the industry has been slow to innovate and adopt offsite technologies, which has inhibited achieving a step improvement in productivity and quality of housing supply. This is all most likely to occur with a combination of government (as potential facilitator, sponsor and major client) and other client and cross-industry support. To date, the level of dialogue and collaboration between designers, contractors and building material producers appears to be limited, and the industry needs to adopt a much more proactive stance on this issue.

The questionnaire survey identified a number of organisational and managerial characteristics that differentiated organisations' attitudes and approaches to sustainable refurbishment where innovative environment technology was implemented. These characteristics also relate to the groupings identified by Rogers (2003) in his 'Diffusion of Innovation' theory. The organisations perceived to be at Level 1 (Initiation Stage) sought to achieve an equal balance between the three sustainability attributes (socio, economic and environmental). Decisions on whether to include sustainability in a refurbishment project were primarily tactical, with little senior management or tenant involvement. No systematic monitoring of the performance of the sustainable refurbishment was undertaken and no long-term strategies or plans existed. This group could easily fall into the category of 'laggards' (Figure 8.1). Organisations who placed themselves at Level 2 (Contagion Stage) placed high importance on the economic and social attributes of sustainability. Management of projects tended to be at the executive level with some monitoring of post project performance being undertaken against embryonic long-term plans. Some tenant involvements in the decision-making process were identified. This category shows signs of 'late majority', who recognise that a shift has occurred in their industry, and make the changes required to adopt the new technology. Those organisations at Level 3 (Control Stage) of the sustainability maturity curve placed high importance on social and environmental attributes of sustainability when making sustainable refurbishment decisions. There are similarities with the 'early majority' category and long-term sustainability plans had been developed and decisions about which projects to pursue were beginning to be devolved to local managers. Organisations that placed themselves at Level 4 (Mature Stage) placed highest importance on the social attribute of sustainability and although they had a formal sustainability policy in place, this tended to be only one of a number of policies that they used to inform their sustainable refurbishment decisions. In interpreting the social needs of their tenants, these landlords demonstrated a community focus with long-term strategic objectives informed by tenant involvement informing all stages of the sustainable refurbishment decision-making process. This category mostly shows signs of 'early-adaptors' moving towards 'innovators' as they aim to attain their sustainable goals. Their vision was strategic and their monitoring was against long-term benchmarks.

8.7 Summary

All case studies illustrate that challenging leadership, organisational structure and management, commitment of all parties, a changed mind-set and better communications all levels are crucial to drive the sustainability actions. Knowledge transfer, sharing of best practice, and training of personnel was essential to make the initiatives successful. From the results of the questionnaire survey it would appear possible to differentiate management characteristics of UK social landlords in their attitudes toward sustainable refurbishment depending upon the level of sustainability at which they perceived their organisations to be. In general, the more advanced level the organisation is perceived to be in sustainability, the more locally focussed the decision making. Further, given the existence of organisational characteristics it should be possible to develop a profiling tool that allows landlords to recognise their position in the sustainability scale and develop appropriate interventions that could accelerate the journey from their current position to a more advanced level. Given the current low level of sustainable housing refurbishment, this could increase the speed of uptake of innovative technologies and support the UK in achieving lower carbon emissions associated with the social housing sector. The existence of barriers, particularly financial, cannot be underestimated and a solution to provide access to funding is needed alongside management interventions. The Green Deal, if effectively applied to social landlords, has the potential to provide such fiscal stimulus and as such, tools to assist with the other aspects of the sustainable refurbishment decision making process need to be developed. In developing these tools consideration should be given to where organisations place themselves in achieving sustainability targets, with solutions being developed that match the characteristics for each level to be achieved. In this way UK social landlords will truly benefit from their investment in Low and Zero Carbon technologies.

Innovation requires skill changes and new technical and practical knowledge. A change in industry, employee culture and mind-set is also essential if the anticipated requirement for improved quality, accuracy and precision is to be achieved successfully. This change in mind-set will require training, an explanation and understanding on the part of the workforce as to why the change is required, and clear leadership at all levels and from all industry participants including clients. The results also identified a range of leadership and organisational characteristics that are needed to drive innovation. It concluded that it is possible to profile UK social housing providers and develop management instruments to accelerate their journey along the sustainable innovation path that in turn will accelerate the uptake of sustainable refurbishment programmes.

References

Ball M. (1999). "Chasing a Snail: Innovation and House building Firms' Strategies", *Housing Studies*. Vol. 14 Issue 1, p9-22.

Barlow, J. (1999). "From craft production to mass customisation: Innovation requirements for the UK house building industry", *Housing Studies*. 14 (1), 23–42.

Broadman, B. (2007). "Home truths: A low carbon strategy to reduce UK housing emissions by 80% by 2050", University of Oxford's Environmental Change Institute, University of Oxford, UK.

Cohen, W. M. and Levinthal, D. A. (1990). "Absorptive capacity: A new perspective on learning and innovation", *Administrative Science Quarterly*, 35, 128–52.

Cohendet, P. and Steinmueller, W. E. (2000). "The codification of knowledge: A conceptual and empirical exploration", *Industrial and Corporate Change*, 9 (2), 195–210.

Department of Communities and Local Government (DCLG) (2012). "The English Housing Survey Homes Report 2010".

Department of Energy and Climate Change (DECC) (2009). "The UK renewable energy strategy", London: The Stationary Office.

Department of Energy and Climate Change (DECC) (2012). "Statistical Release: 2011 UK Greenhouse Gas Emissions, Provisional Figures and 2010 UK Greenhouse Gas Emissions, Final Figures by Fuel Type and End-User", online, at: http://www.decc.gov.uk/en/content/cms/news/pn12_033/pn12_033.aspx.

Department of Trade and Industry (DTI) (2003). "Energy white paper: Our energy future – Creating a low carbon economy", Norwich: The Stationary Office.

Department of Trade and Industry (DTI) (2007). "Meeting the energy challenge: A white paper on energy", Norwich: The Stationary Office.

Earl, M. (1989). *Management strategies for information technology*. Upper Saddle River, NJ: Prentice Hall.

Gann, D (2001). "Putting academic ideas into practice: Technological progress and the absorptive capacity of construction organizations", *Construction Management and Economics*, 19 (3), 321–30.

Gaterell, M, and McEvoy, M. E. (2005). "The impact of climate change uncertainties on the performance of energy efficiency measures applied to dwellings", *Energy and Buildings*, 37 (9), 982–95.

Hinks, J., Alexander, M., and Dunlop, G. (2007). "Translating military experiences of managing innovation and innovativeness into FM", *Journal of Facilities Management*, 5 (4), 226–42.

Hooper, A. J. and Nicol, C (2000). "Design practice and volume production in speculative house building", *Construction Management and Economics*, 18, 295–10.

Kaluarachchi, Y. & Jones, K. (2007a). "Promoting innovative technologies in the housing sector in the UK", *Construction Management and Economics*, 25th Anniversary Conference proceedings, University of Reading, UK.

Kaluarachchi, Y. and Jones, K., (2007b). "Monitoring of a strategic partnering process – the Amphion Experience", *Construction Management and Economics*, 25 (10), 1053.

Lawson B., and Samson, D. (2001). "Developing innovation capability in organisations: A dynamic capabilities approach", *International Journal of Innovation Management*, 5 (3), 377–400.

Pan, W., Gibb, A. & Dainty, A. (2005). "Offsite modern methods of construction in house building perspectives and practices of leading UK house builders", Loughborough University.

Power, A. (2006) "Stock Take: Delivering improvements in existing housing", London: The Stationery Office. *Waste and Resource Management*, 2006; 159(2):65e72.

Rogers, E. M. (2003). *Diffusion of innovations* (5th edition). New York, NY: Free Press.

SAMI Consulting, (2008) 2020 Vision – The Future of UK Construction St Andrews Management Institute 83.

Seaden, G., Guolla, M., Doutriaux, J. and Nash, J. (2003). "Strategic decisions and innovation in construction firms", *Construction Management and Economics*, 21, 603–12.

Sustainable Development Commission (SDC) (2005). "Sustainable buildings: The challenge of the existing stock", London.
Sustainable Development Commission (SDC) (2006) "'Stock Take': Delivering improvements in existing housing", London.
WWF (2008). "How low: Achieving optimal carbon savings from the UK's existing housing stock", WWF-UK.

9 Leadership studies in changing times

Leighton A. Ellis, Timothy M. Lewis, and Andrew K. Petersen

9.1 Introduction

In order to ensure that the leaders of the future are properly prepared, the educators of today need to rationalise what this means in terms of the skill set that they will need. There is an increasing need for this preparation to include an ability to address sustainable development in terms of the economic, technologic and socio-cultural issues that may be involved. Universities need to take the lead in this global initiative by defining the skill set, through defined learning outcomes that will enable our future engineers to create that sustainable future. One university has sought to achieve this through a total restructuring of its civil engineering programmes in order to reform the type of graduates being produced, including attempting to improve their leadership capabilities. The restructuring involved not only a complete review of all existing courses, but also the introduction of new ones to cover areas of perceived weakness, as well as a complete overhaul of the methods of assessment being used. Again, in this latter process, new techniques were introduced as the existing techniques tended to assess the wrong things or assess them in the wrong way. The new assessment techniques that were introduced included one defined as '*Zero Tolerance*' and another which was based on '*360-Degree Feedback*' which were designed to measure the students' learning as well as their grasp of sustainability concepts.

While all courses were updated, one particular new course was aimed specifically at finding a way to assess 'leadership' amongst other things. Most university programmes either do not attempt to assess leadership, or test it in terms of what students recall or understand about leadership theory. The new course runs for one full semester of fifteen (15) weeks, and it is known as the *Practical Team Project*. The course is structured around group projects on which a team of students '*identify and solve technical, business, social, cultural and ethical issues for a given Project both systematically and creatively, and make sound judgments in the absence of complete data*' (UWI Student Handbook, 2012). The projects are based on real project situations in the local environment, and must exhibit environmental awareness and knowledge of sustainability issues, as well as requiring each student to demonstrate his or her leadership capabilities. The course has proved not only popular, but also the methods of assessment that were introduced to assess 'demonstrated leadership' have proved to be consistent and able to produce a measure of 'leadership'.

9.2 Background

In the existing literature, the most commonly cited definition of sustainable development is that of the 1987 Report of the UN World Commission on Environment and Development, commonly known as The Brundtland Report, which contained the statement (WCED, 1987:8):

> *'Humanity has the ability to make development sustainable – to ensure that it meets the needs of the present without compromising the ability of future generations to meet their own needs'.*

In various international conferences since then governments have come out with statements, strategies and commitments intended to express their wholehearted adoption of the principle of sustainability (OECD, 2001). In July 2006, the Institution of Civil Engineers (ICE) along with the American Society of Civil Engineers (ASCE) and the Canadian Society for Civil Engineering (CSCE) signed an agreement on sustainable development (ASCE, 2006), with each society pledging to implement action plans (ASCE, 2009). In July 2007, the (ICE), the Association for Consultancy and Engineering (ACE), the Civil Engineering Contractors Association (CECA), the Construction Industry Research and Information Association (CIRIA) and the Construction Products Association, published a report (ICE, 2007) that set out the terms under which they would cooperate in trying to achieve sustainability. This has been successful in a number of ways including the decision to require the inclusion of sustainable development principles into undergraduate degree courses accredited by the ICE, and into the professional qualifications process. In addition, the associations aim to promote leadership, continuing education and training, knowledge transfer, support for research and innovation and a stronger 'voice' for the industry and profession to government and the public.

The ICE, along with three other professional bodies, the Institution of Structural Engineers, the Institution of Highways and Transportation, and the Institute of Highway Incorporated Engineers, have formed the Joint Board of Moderators (JBM) to work and strengthen links with universities to ensure that appropriate educational programmes are in place to provide the skill set needed by professional engineers. This includes a commitment to trying to maintain sustainable economic growth and necessary ethical standards (JBM, 2010). The JBM assesses and makes recommendations on the content of particular educational programmes. If these are acceptable they will be accredited as meeting the requirements for registration as a professional engineer with the Engineering Council (UK) – this is the body responsible for regulating the engineering profession within the UK (JBM, 2010).

9.3 The restructuring of the civil engineering programmes

In the later 1990s, the Dearing Report (Dearing, 1997) had refocused universities in the UK on the need for a 'programme specification' rather than a 'teaching syllabus' to be prepared for each course, with this specification being built around measurable learning outcomes. Measurable means that each learning outcome must be assessed by either coursework or examination, with students having to demonstrate their competence, up to the minimum level of a pass, at each. This of course meant that examinations could no longer be to answer 'any five from eight' because every outcome had to be assessed and had to achieve a pass mark – so it became 'answer all questions'. And the questions had specifically to cover the range of learning outcomes.

Towards the end of 2006, the Department of Civil and Environmental Engineering at the University of the West Indies (the Department), decided that a major review of its programmes should take place. In a previous JBM Accreditation Report the Department was asked to demonstrate, through a paper submission, amongst other things, how the learning outcomes are assessed and the range of industrial involvement. The concept of learning outcomes was unfamiliar to the staff, because until this time, the teaching of the Department had been focused on 'teaching objectives' rather than 'learning outcomes'. Lecture schedules were defined in terms of at least one teaching objective per lecture. It was not only this Department that operated in this manner, because none of the programmes anywhere in the university were directed at learning outcomes. As this was the only university in the country at the time, staffs were not exposed to the concept by colleagues elsewhere except at conferences or University visits overseas. Thus it was a lonely path that was taken. Fortunately a guide along this path was available in the form of a lecturer from the UK who had recently joined the university and was very familiar with the whole concept behind 'learning outcomes'. He proved to be a critical and invaluable resource, and it is unlikely that the task would have been accomplished without his input.

9.3.1 A Prescription

In 2006 a new strategic plan (2007-2012) was put forward for the University of the West Indies, by the Vice Chancellor, Professor E. Nigel Harris, which set out a description of the desired attributes of all graduates of the university. He wrote that they:

> *'should be competent and knowledgeable, leaders and team players, critical and creative thinkers, effective communicators, problem solvers,*

> *IT and information literate, socially and culturally responsive, ethical, innovative, entrepreneurial and life-long learners'*, (UWI, 2007).

Although these had always been desirable characteristics they now became a model for what was required. In response the Department rewrote its mission statement along similar lines as being

> *'To provide quality education at undergraduate and post graduate level in order to produce critical and creative thinkers, problem solvers, effective communicators, knowledgeable and, competent leaders, team players, IT and information literate, socially and culturally responsive, ethical, innovative, entrepreneurial and lifelong learning engineering graduates, who will enhance the quality of life and face future challenges within the Caribbean and beyon'.*

The wording was a bit clumsy, but the intent there was to use the Vice Chancellor's plan as not just a description but a prescription for the learning outcomes of future programmes. Hence, the Department identified that on completion of the MSc Civil Engineering a successful candidates would be able, at the threshold (pass) level, to:

1. *Lead and work within teams* to identify and *solve technical, business, social, cultural and ethical issues* in Civil Engineering both systematically and *creatively,* make sound judgments in the absence of complete data, and *communicate* their conclusions clearly to specialist and non-specialist audiences;
2. Demonstrate self-direction, *critical thinking* and *originality* in tackling and *solving problems*, and act autonomously in planning and implementing tasks using *Information Technology.*
3. Continue to advance their *knowledge* and understanding, and to develop their new skills to a higher level. Candidates will have the *competencies, qualities and transferable skills necessary for employment requiring:* the exercise of initiative and personal responsibility; decision-making in complex and unpredictable situations; and the independent *learning ability required for continuing professional development* as a practicing Civil Engineer.

The review of the programmes in the Department had to ensure that these learning outcomes would be achieved mainly at the undergraduate level, but some after a postgraduate programme. Most of these topics had traditionally been covered in lecture materials, but with a different teaching emphasis. Under the new dispensation it had become necessary to assess, in a measurable way, a number of outcomes that had never been assessed before – such as the student's ability to work within teams or to lead teams. It became necessary to consider how to get students to demonstrate leadership or critical thinking and then to find measurable ways to assess such things. Although universities elsewhere had been embarked on the learning outcome path for over a decade by this time, the literature was light on many of these issues, perhaps because other universities were determined on a different skill set for the graduates. What was immediately obvious was that the students would have to actually work in groups, lead a group, undertake a design oriented project and undertake a research project, in addition to and supported by the normal taught units. The demands of the profession also had to be met and this increasingly focused on the need for young engineers to be fully aware of the need for attention to health and safety issues as well as to the need to try to design for sustainability and to fully appreciate the ethical responsibilities of a professional. It was desirable that these three threads should run continuously through every course in the programme – to the extent of being brought into every lecture particularly in the post-graduate programmes.

9.3.2 Threading sustainability throughout the MSc programmes

The focus here is on the postgraduate programmes, and particularly on the M.Sc. in Construction Management, because it was being put up for accreditation by the JBM for the first time and so it had to receive particular attention. The content of the programme was quite dramatically revised in order to address the prescriptions of the Vice Chancellor and of the profession. The programme had undergone a number of changes in the past. It had started out structured around six full academic year courses with examinations at the end, followed by a research project. The courses also required two major written assignments to be handed in, one each term. When the university semesterised, the programme was restructured to involve twelve courses, six each semester, each with one written assignment and exams at the end of each semester all followed by a research project. These twelve courses represented roughly half of the content of the six courses that they replaced.

While this seemed satisfying at first, it soon became apparent that there was significant grade creep in that the student were getting higher marks than they had previously, with no apparent change in the quality of the students or in the teaching, except for the fact that students had smaller 'chunks' of knowledge to absorb before being examined on them. At this stage each lecture period had been reduced from the two hour duration when there were six courses, to one hour apiece. Lecturers began to feel that this was not enough to develop understanding of the issues being raised, and with students often strolling in late, the relatively short lecture was compromised even more.

At this stage the University decided that all postgraduate programmes should be conducted in the evenings, so that students who were employed could attend after work. This led to further review of the programme resulting in it being reduced to eight courses, one each evening from Monday to Thursday (with Friday kept free – for obvious 'social' reasons). There were four courses each semester which were examined at the end of that semester, followed by the research project. The lecture time was increased to two hours per course, normally running from four in the afternoon to six, though some optional courses ran from six to eight. The contents of these courses were derived from the twelve that preceded them. There was no real logic behind the way the contents were shared out; this largely resulted from the academic interests of the lecturers involved. This did ensure that the lecturers involved were generally knowledgeable and interested in the subjects they were covering. The information booklets carried a listing of the curriculum for each course, and the lecturers generally produced a listing of topics to be covered in each lecture – this effectively set out their teaching objectives, listing some twelve topics per course. Examinations could cover any aspect of any topic covered, and students would normally be required to answer and five from eight.

When the need arose to set out learning outcomes and to cover the specific skill set required by the profession, and to attest to the fact that the students were competent in those skills, it became apparent that another change was required. This review and restructuring was rather more well-thought-out than previous ones. Key structural changes now meant that the courses were developed around the learning outcomes, that the students now knew precisely what skills they were required to have and to be examined in, so they did not need to study a whole set of interesting but extraneous details. As a result, also, exams would require every question to address a learning outcome and for all to be answered. A student would be required to achieve a pass mark on each question, so that they could be said to be competent, at the threshold level, in each.

The restructuring in this instance required certain courses to be dropped altogether because they did not have relevant learning outcomes, or for the relevant parts of their content to be woven into other courses. This resulted, for example, in a course on 'Construction Economics' being dropped from the programme, while parts of its content (such as an understanding of economic sustainability and cost-benefit analysis for example) were absorbed by other courses. When this exercise was complete, it became apparent that some skills were either not being covered or not being examined. This effectively meant that new courses had to be introduced to capture the learning outcomes that were not being covered otherwise, or which did not enable appropriate forms of assessment – hence the development of the innovative 'Practical Team Project', and the introduction of a course in 'Research Methods' for example. The latter course was required because the university recognised that the M.Sc. is universally considered a research degree, and therefore had to include some element of learning about research methods, as well as the 'research project'.

Furthermore, the profession required that students should have a thorough understanding of sustainability, health and safety issues and professional ethics. The JBM interpreted this to mean that a degree programme should attempt to thread these subjects throughout the programme. This meant trying to bring sustainability into each course in the programme, and so it was necessary to understand the different dimensions of sustainability.

Sustainability and sustainable development are commonly conceptualised as having three dimensions: environmental, social and economic (Academy of Engineering, 2005). Given these different dimensions it became easier to see how to thread sustainability into the M.Sc. programme. Table 1 below shows how each course links to a dimension of sustainability and how these links feed into the desired learning outcomes. Although the M.Sc. in Construction Management has been used as an example, all the other M.Sc. programmes are similarly structured and similarly threaded, though with different taught content.

It can be seen from Table 9.1 how specific learning outcomes were allocated amongst the courses, and how at least one dimension of sustainability was taught and assessed, either through an exam or coursework, in each course. in the one particular course that we are focusing on here, the Practical Team Project, the students, working in teams, are required to draw on all the knowledge gained from the other courses to practically complete a Project Plan. the learning outcomes are defined and the dimensions of sustainability associated with each are shown here.

9.3.3 Assessment of leadership

One perplexing issue that immediately became apparent from this analysis, and defined in the first learning outcome listed, was how to assess a student's demonstrated contribution to team work and their demonstrated leadership capabilities. Clearly this required each student to take a leadership role and have his or her performance evaluated on some basis. Given the restrictions of a conventional M.Sc degree programme's time constraints this presented a new problem. The solution came out of one young member of staff's deep interest in leadership issues, and the requirements of the ICE for becoming a Chartered Engineer. After numerous discussions and iterations of ideas, the design of the Practical Team Project began to emerge.

First it was decided that the Practical Team Project should have four learning outcomes which pick up issues/skills not covered in other courses and which should be aligned with the dimensions of sustainability. One such issue was the need for the students to lead and to work in teams; ideally, for this to be meaningful the students should also undertake this team-work on realistic projects. The result, naturally, was for the course to be structured around current projects in the local environment, with the development of a project implementation plan being the final output. Along the way, each student should be required to take turns in leading the team while finding solutions to all of the technical, business, social, cultural and ethical issues that they encounter. Because the projects were based on real world problems, the students would be required to be systematic and creative in finding solutions as well as to show that they could make sound judgement in the absence of complete information.

Table 9.1 Showing the threading of sustainability in the UWI M.Sc. Construction Management program

MSc Courses (Construction Management)	*Learning Outcomes*				
COEM 60XX - Natural Hazards Management	Identify and evaluate the threats posed by natural hazards in the Caribbean, (Eco-centric)	Apply and justify approaches to mitigating threats to human life and the environment. (Socio and Eco- centric)	Apply and justify approaches to maintaining essential services. (Techno-centric)		
COEM 60XX - Construction Accounting & Finance	Evaluate Financial Accounts (Techno-centric).	Evaluate Management Accounts (Techno-centric).	Analyze the economic, environmental and social impacts of a project during its life cycle (Socio, Eco, and Techno-centric).		
COEM 60XX - Organisation & Management of Construction	Identify and evaluate the role of construction in the national economy (Eco-centric).	Compare and contrast different managerial styles and organisational structures, and identify effective leadership styles (Socio-centric).	Evaluate the impact of risk on decision making, and apply techniques that can enhance creativity (Techno-centric).	Know and apply the legal requirements regarding health and safety on site (Socio- centric).	Evaluate the need for quality management systems in a firm or on a project (Eco-centric).

COEM 60XX - Materials Technology	Identify and evaluate the properties and behaviour of materials for use in construction. (Teehno-centric)	Identify the factors affecting the production of construction materials and the patterns of demand in the Caribbean. (Eco and Techno-centric)	Understand the use of specifications, standards and testing of materials (Techno-centric)	Demonstrate understanding of the basic principles of materials procurement, handling, storage (Socio-centric)
COEM 60XX - Contract Management & Construction Law	Identify and evaluate the contract options available for the procurementof goods and services(Socio-centric).	Apply and justify contract and risk management legal procedures (Eco- centric).	Compare and contrast traditional and modern contract management and legal relationships (Techno-centric).	Negotiate a settlement of a contractual dispute (Socio-centric).
COEM 60XX - Construction Project Management	Identify and explain the structure and phases of a construction project (Techno-centric).	Evaluate and apply activity scheduling to a construction project (Eco-centric).	Evaluate the costs associated with elements of work on a construction project for effective cost management (Eco-centric).	Identify and evaluate the health and safety requirements of a project (Socio and Eco-centric).

(Continued)

Table 9.1 (Continued)

MSc Courses (Construction Management)	*Learning Outcomes*			
COEM 60XX - Maintenance & Facilities Management	Identify and evaluate the maintenance necessary for particular materials, elements of structures and complete projects(Eco and Techno-centric).	Establish a maintenance policy and justify a maintenance budget (Techno-centric).	Undertake a maintenance survey and performance evaluation (Eco and Techno-centric).	
COEM 60XX - Research Methods	Formulate a research proposal on a chosen topic(Techno-centric).	Explain the principles of research within the context of a chosen topic (Socio, Eco, Techno-centric).	Justify the research design in a research proposal (Techno-centric).	Justify the research methodology in a research proposal (Techno-centric).
COEM 60XX - Practical Team Project	Lead and work within a teams (Techno-centric).	Identify technical, business, social, cultural and ethical issues for a given Project (Techno, Socio and Eco centric).	Solve technical, business, social, cultural and ethical issues for a given Project both systematically and creatively, make sound judgments in the absence of complete data (Techno, Socio and Eco-centric)	Communicate their conclusions clearly to specialist and non- specialist audiences (Techno-centric);

	Team 1					Team 2					Team 3				
Objective	1	2	3	4	5	6	7	8	9	10	11	12	13	14	15
Project Scope Management															
Project Risk Management															
Project Time Management															
Project Cost Management															
Project HR Management															
Project Quality Management															
Project Procurement Management															
Project Communication Management															
Project Integration Management															
Submission of Project Plan															
	L	TP	TP	TP	TP	L	TP	TP	TP	TP	L	TP	TP	TP	TP

L Leader
TP Team player

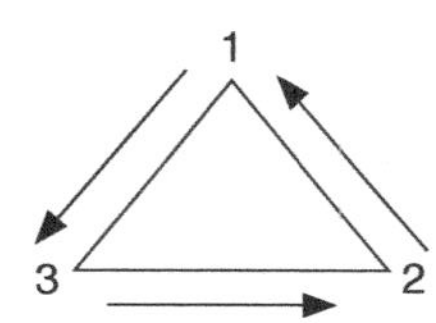

Figure 9.1 Rotational diagram of leaders in teams

(Sourced from Ellis & Petersen, 2009)

The approach adopted to force the students to go through the full process of plan development was to refer to the Project Management Book of Knowledge (PMBoK, 2004), and to structure the weekly sessions of the course around the knowledge areas of the PMBoK. These knowledge areas describe things that have to be managed to successfully complete a project, and are listed as the objectives at the left of Figure 9.1. Each individual student is listed along the top grouped into teams, and the coloured cells indicate how the different objectives are addressed by the different students as time passes – the student's progression is charted by a column in the table. The row at the bottom of the table indicates who is the leader of each team, and who are the team players. Without making it obvious, health and safety, sustainability and professional ethics were expected to be addressed within each objective.

Again referring to Table 9.1, in team 1, individual 1 follows the black cell pathway starting in Team 1 as leader, but then moving to Team 3 as a Team Player, and continuing in Team 3 until week six when he/she becomes leader. In week 7 that individual moves into Team 2 as a Team Player, and so on.

This fifteen (15) week course is delivered in the final taught semester of the program; the nine (9) knowledge areas of the PMBoK were used as weekly objectives assigned to the leaders of the various teams. The teams consist of five (5) members – one leader and four team players (see Figure 9.1). The leader is required to divide the workload required to accomplish the objective for that week amongst the team players and motivate them to achieve the objective before the commencement of the class the following week. The leader is then assessed for demonstrated leadership by the client on the project using a questionnaire to assess his or her effectiveness in achieving a complete and acceptable outcome on the set objective. The leader's performance is thus measured by the team's output, on the basis that a more effective leader will better motivate the team to perform. Each leader is given the opportunity to lead a team on two occasions, This should help to even out some of the variation between having to lead in the first week without experience of the responsibility, and leading after five weeks when there is significant experience to learn from.

In order to assess the contribution of each individual student to the work of the team a technique relatively new to academia was introduced. Human Resource Management has for some time been using a technique called 360-degree feedback. This typically requires the individual's performance to be assessed by the people above, below and at the same level, as well as the individual him or herself. In this instance the students do not really have anyone 'below' them so the bespoke model used here simply considers the opinions of the lecturer/project client, the other team members and themselves (see Figure 9.2) – this was reported on elsewhere by Ellis & Petersen (2009). Although likely to be biased and subjective, self and peer assessments have proved capable of providing summative feedback (Mowl, 1996; McDowell and Mowl, 1996), which can be used to assign a mark or grade to a student.

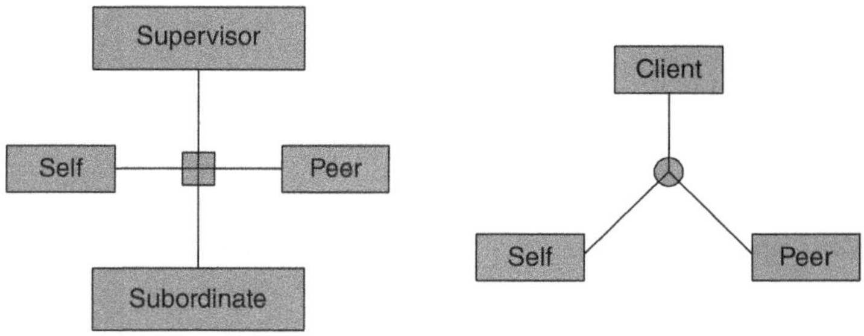

Figure 9.2 Traditional (left) and Bespoke (right) 360-Degree Feedback relationships

An objective means of assessment was also included to assess the quality of work produced by the team players. Reynolds *et al.* (2004) developed an approach based on zero tolerance to errors and omissions which was adopted here. Under this Zero Tolerance approach, the team players were assessed on their ability to produce a one page document without errors and omissions based on the team objective and the specific task assigned by the team leader. If more than three errors or omissions (including spelling, punctuation and grammar) were noted in this document, three were marked and the document was returned to the team member to correct and resubmit. The students were each given five (5) opportunities to submit an error free document each week – each resubmission involved a loss of marks both for the team member and the leader. The leader was, of course, allowed to review every document, and try to get it correct, before it was submitted. This was found to provide a reliable basis for assessing students' performance as it tended to mirror their academic achievements. The students' performance on average showed the number correct at the 1st attempt = 80 percent 2nd attempt 70 percent of remainder, 3rd attempt 60 percent, 4th attempt 50 percent, 5th attempt 40 percent – any who did not achieve an error free document after that were deemed to have failed. In this project-based learning environment, intended to throw up situations that reflect the real world, the students are expected to bring into their project-oriented thought processes the implications of sustainable development, health and safety, risk management, design implications and professionalism. As part of Project Integration Management, the students are required to show how they have threaded sustainability, health and safety etc. throughout the project.

9.3.4 Analysis of Assessment Methods

It was hoped that the different methods of assessing student performance used in this course would give consistent results. If this happened it would tend to suggest that each individually was consistent and meaningful, and that taken together they would verify and validate one another. To examine the results of using the different techniques, the Statistical Package for the Social Sciences (SPSS, 2009) software was used. The detailed results are reported on elsewhere.

It is of interest to note on the results of the 360 degree feedback instrument, that from the results so far (the cohorts from three academic years) there is no significant difference between the assessments given by the peers, the lecturer and the student team-leaders themselves. This is rather surprising as individuals are not normally particularly modest in their own assessment of their performance, but it may be due to the fact that this was a new programme and they wanted to be as accurate as they could, knowing that it would be subject to careful analysis later. This outcome is actually very desirable, because it shows that the people can be objective and agree with one another on a subject that is largely subjective, and regardless of their status-relationship with one another. In other words, we know the quality of leadership when we see it in action, whether we are the supervisor, the colleague or in the position of leader personally. A detailed analysis of the Zero Tolerance exercise was that students tended to take between two and three attempts to get it right – actually on average around 2.7 attempts. A normal distribution would suggest that a mark of close to three should be obtained. This tends to confirm that the assessment procedure is being implemented effectively and accurately. It was noticed that as the exercise progressed the performance improved, going from over three to under three attempts to get it right. It seems that the students became more aware of their ability to improve their marks by taking a little extra care, or by getting a colleague or the leader to also read through their work and to correct it accordingly.

It is felt that the fact that students received almost immediate feedback (the same week) on most of the exercises in this course played a major role in helping them to improve their performance. It is unusual for students to know so quickly how they are doing and where they are going wrong. This was a key component of the course design. A conscious effort has been made to get successful students to continue their involvement with the department and their interest in the programme by persuading them to return after graduation to act as clients for new projects in the course. This policy has been successful and the fact that the clients are themselves recent participants in the course means that they are well equipped to know what is needed from a client and what is expected from a student when acting either as leader or team player.

9.4 Summary

The profile expected of a new graduate from an engineering postgraduate programme has changed and will continue to change. It is important that the academic training they get is kept up-to-date. The resulting changes in the skill set may mean that the learning outcomes change and that they need new and innovative methods of assessment. This chapter describes a case study of the process of changing a degree programme at university in the difficult position of being the only university in the country with Civil Engineering at the time. The lack of other universities 'down-the-road' makes change even more difficult in a traditionally conservative institution. Key factors which enabled the change to take place were the desire to meet accreditation and professional requirements and to meet the directives of the Vice Chancellor, as well as the presence of a change champion in a position to make decisions, and a key knowledge gatekeeper who enabled the necessary information on the logic behind the learning outcome approach to teaching to be available, as well as the implications of the JBM and ICE skill sets in the accreditation process, and who then took the lead in implementation.

The secondary case study was that of the development of a new course that could meet the need to assess student performance in areas that are not normally measured – such as team work and leadership for example. This was possible because of the commitment and interest of another knowledge gatekeeper who was informed on the ICE requirements for professional membership as well as the issues surrounding the assessment of leadership. It is clear that these three resources were all essential and were fortunately all available when needed. The initial findings have been that the newly structured programme is proving very effective and popular with the students. There has been no significant creep, either up or down in the students' marks, so the overall effect has not handicapped nor advantaged the students involved. Completion rate has remained as before. For the particular new course described here, the Practical Team Project, the marks that are being obtained appear to be consistent with one another and to verify and validate one another. For example it was surprising but satisfying that the supervisors, peers and individuals themselves all gave consistent grades for the individual's performance. This suggests that the measures are valid. Similarly the Zero Tolerance exercise gave results that were very close to what would be expected from a normally distributed grading, which again serves to help verify and validate the methodology.

References

American Society of Civil Engineers (ASCE) (2006). "Developing Leadership". Annual Report 2006. Available at: http://content.asce.org/annualreport/2006/2006.pdf [Accessed 1 March 2010]

American Society of Civil Engineers (ASCE) (2009). "Sustainability Action Plan". Task Committee on Sustainable Design. Available at: http://content.asce.org/Sustainability/ActionPlan.html [Accessed 1 March 2010].

Dearing, R. (1997). "Report of the National Committee of Inquiry into Higher Education". Available from: http://www.cosmosft.demon.co.uk [accessed 17 July 2001]

Drucker, P. (1954). The Practice of Management". New York: HarperCollins Publishers Inc.

Ellis, L. and Petersen, A. K. (2009). "Leadership traits: An assessment of post-graduate students at the University of the West Indies, Trinidad and Tobago". Proceedings of International Association of Education Assessment Conference, Brisbane, Australia.

Institution of Civil Engineers (ICE) (2007). Sustainable Development Strategy and Action Plan for Civil Engineering. Available at: http://www.ice.org.uk/Information-resources/Document Library?categoryid=36&page=12, [Accessed 2 January 2014)

Joint Board of Moderators (2010). "Welcome to JBM". Available at: http://www.jbm.org.uk/index.aspx, [Accessed 9 March 2010], Joint Board of Moderators (JBM)

McDowell, L. and Mowl, G. (1996). "Innovative assessment – its impact on students". In Gibbs, G. (ed.) *Improving Student Learning Through Assessment and Evaluation*. Oxford: The Oxford Centre for Staff Development, 131–47,

Meadows, Donella H., Dennis L. Meadows, Jorgen Randers, and William W. Behrens III.

(1972). "The Limits to Growth". Washington, D.C.: Potomac Associates, New American Library.

Mowl, G. (1996). "Innovative Assessment". in DeLiberations. Available at: http://www.lgu.ac.uk/deliberations/assessment/mowl_content.html, [Accessed 31 March 2009]

Organisation for Economic Co-Operation and Development (OECD) (2001). *The DAC Guidelines. Strategies for Sustainable Development: Guidance for Development Co-operation*. Available at: http://www.oecd.org/dataoecd/34/10/2669958.pdf [Accessed 28 February 2010]

Pepper, C. (2007). "Sustainable Development Strategy and Action Plan for Civil Engineering". Institution of Civil Engineers, One Great George Street, London SW1P 3AA.

Perks, A., Burrell, B., Korol, B., Khan, A., Heroux, J., and Ford, L. (2007)

"Entrusted to Our Care – Guidelines for Sustainable Development". The Canadian Society for Civil Engineering. Available from: http://www.csce.ca/docs/CSCE-SD%20Guidelines_22Jan07.pdf [Accessed 9 March 2010].

PMBoK (2004). *A Guide to the Project Management Book of Knowledge: PMBoK Guide*. 3rd Edition. Pennsylvania: Project Management Institute Inc.

Reynolds, J. H., Petersen, A. K., andTutesigensi, A. (2004). "Evaluation of a Zero Tolerance Assessment Strategy for Incorporating Risk Assessment into Undergraduate Construction Related Courses" (LTSN Engineering Mini-Project Report). In LTSN Engineering (ed.) *Mini-Project Report* (pp. 1–10). Loughborough: LTSN Engineering.

The Royal Academy of Engineering (2005). "Engineering for Sustainable Development: Guiding Principles". http://www.raeng.org.uk/events/pdf/Engineering_for_Sustainable_Development.pdf [Accessed 28 February 2010]

SPSS 2009. *Statistical Package for Social Scientists*. Available at: http://www.spss.com/ [accessed on 15th July 2009]

UWI (2007). "UWI Strategic Plan 2007-2012". Office of Planning and Development. Available from: http://www.uwi.edu/StrategicPlan/ strategicplan20072012.aspx, [Accessed 20 July 2009]

World Commission on Environment and Development (WCED) (1987). *Our Common Future (The Brundtland Report)*. Oxford: Oxford University Press.

Index